Asking the E...

WININ PEREIRA and JEREMY SEABROOK

Earthscan Publications, London

First published 1990 by
Earthscan Publications Ltd
3 Endsleigh Street, London WC1H ODD

Copyright © Winin Pereira and Jeremy Seabrook 1990

All rights reserved

British Library Cataloguing in Publication Data
Pereira, Winin
    Asking the earth
    1. India (Republic). Social conditions.
    I. Title. II. Seabrook, Jeremy
    954.052
ISBN I-85383-045-3

Designed by Sandie Boccacci
Production by Bob Towell
Typeset in 10/11½pt Century Old Style by Selectmove, London
Printed and bound by Guernsey Press Ltd

Earthscan Publications Ltd is a wholly owned and
editorially independent subsidiary of the International
Institute for Environment and Development.

Contents

	Acknowledgements	iv
	Foreword	v
	Introduction	ix
1	The other side of history	1
2	The sustainable lifestyle of the Warlis	27
3	Red ink in the "Blueprint for Survival"	61
4	Technological intervention	89
5	Farming systems	109
6	Interconnections of violence	130
7	Sa vidya ya vimiktaye!	144
8	The bhagat and the allopath	153
9	Natural versus formal forestry	182
10	Celebrating trees, celebrating life	193
11	Restoring our future	213
	Glossary of plant names	221
	Index	224

Acknowledgements

I should like to acknowledge the contribution to this book of the Adivasis of Thane district, in particular Devki Markari, Basanti Dandekar, Nirmal Dumada, Janiya Ghatal, Lahanu Rade, Gopal Rade, Lahu Dongarkar and Babu Shanwar. Others who have helped tremendously are Vasudha Ambiye, Shanthi Kurien, Anu Gupta, John Dabre, Manek Mistry, Celine D'Souza, Maya Apte, Jeeten Bhat, the members of the Maharashtra Prabodhan Seva Mandal and the members of my family, particularly Ravi and Nikhil. In the collection of people's knowledge, it is not possible to single out the countless individuals whose lives exemplify a hopeful traditional wisdom.

My special thanks are due to Jeremy Seabrook, without whose encouragement and suggestions, this book would not have been produced. It has been a pleasure working with him. For both of us, it has been revealing to discover how far the interests of people of both North and South coincide.

While care has been taken to acknowledge sources of ideas, it is impossible to mention the originators of all of them, and I acknowledge my indebtedness to those whose names have been omitted.

I believe ideas should be free and would welcome others to use any they might find worthwhile, with or without acknowledgement.

WININ PEREIRA
January 1990

Foreword: The Spread of Unsustainable Development

The unequalled power and wealth of the West have ensured that patterns of development, originally imposed upon the rest of the world through colonial conquest, can now proceed under a cloak of economic necessity. The very word "development" has come to have a deterministic ring – it means the path pursued by the West. What it signifies in practice, and in the experience of people, is growing market penetration of increasing areas of human life. It means the enclosure of common resources, and the transformation of these into commodities and services that can be acquired only through monetary transactions. This is a powerful and dynamic process, sanctioned and mediated by the "science" of economics; although it is becoming increasingly plain that existing economic systems have no measure for the way in which they themselves affect the natural resource-base of the earth.

In the countries of the Two-Thirds World (for two-thirds of humanity live in what is commonly misnamed the Third World), where people have remained close to the resource-base upon which they depend, this has long been apparent; only now is the West beginning to realize that the natural world is neither a limitless provider of raw materials, nor an infinite absorber of all the noxious by-products of industrialism. Because the countries of the Two-Thirds World have in the main been the furnishers of primary commodities, it is they who have borne the brunt of resource-depletion, deforestation and lowered water tables – until now, the West has serenely continued to "develop", apparently oblivious to the malign consequences of its freedoms.

The process of appropriating and enclosing common resources for the benefit of the market economy, and the individuals who control and direct it, occurred long ago in the West; and was sustainable for as long as the resources of the rest of the world could be compelled into its service.

Most of the Two-Thirds World had sustainable systems in place until the advent of colonialism. Colonialism should perhaps be seen as the means which enabled Europe to nurture its own unsustainable system and to destroy its own natural capital. The spread of neo-colonial Western development continues the process of despoilation. A retrieval of sustainable practices is the most urgent task facing humanity.

This book reflects the experience of someone who accepted all the orthodox nostrums of development that Western-inspired interests have foisted upon the Two-Thirds World for the past 30 years, and who learned from direct practice that these were both harmful and impoverishing. It is not easy to disengage from damaging patterns of growth that pass on their hidden costs to the poorest and least defended: at every turn, the effort to find another way strikes at the vested interests of those who are agents of alien forms of economic growth.

The book is addressed both to people in India and the West: it shows that it is impossible for the countries of the Two-Thirds World to liberate themselves from harmful forms of development while the West continues as before. For the impoverished have been appointed, as always, to resolve the contradiction of a ruinous western industrial society. Sustainable development is all the rage now, but the version of sustainability propounded by the West (and exemplified by the Pearce report), is seen as an affront to those whose whole lives have been dedicated to the same idea from the vantage-point not of the rich, but of the poorest on earth. It is hypocrisy to preach freedom, and then to pass on the costs of those freedoms to the poor and to their threatened habitat. This is why Chapter 3 is concerned with a clarification of the contradictions within the Pearce report.

The book is intended to stimulate and widen discussion of what we believe to be some of the most vital issues facing humanity. We shall be happy to engage in constructive dialogue with any groups or individuals who care to respond.

It has been a great privilege to work with Winin Pereira, and I have been happy to help in the editing of these papers, which have appeared as articles in *Anusandan*, a periodical of Maharashtra

Prabodhan Seva Mandal, in recent years. Any contribution I have made is offered with respect and affection.

JEREMY SEABROOK
Bombay, 1990

Introduction

Winin Pereira was born in 1928, in Bombay. He was a physicist, but gave up his work in 1959 because of his disillusionment with science and in order to spend more time on the problems of social justice. He was one of the founding members of the Maharashtra Prabodhan Seva Mandal, which was formed with the intention of helping the poorest farmers.

Since that time, Winin has monitored the changing fashions of Western-inspired development, and the ways in which even the best of intentions often end up by benefiting the well-to-do, at the expense both of the poor and of the environment.

Originally, the project was concerned with improving the agriculture of small farmers in Nasik district of Maharashtra State, of which Bombay is the capital. Surface-well irrigation and small tractors were the main inputs. Only locally collected funds were used. Young people came from Bombay to work with the Mandal, many for idealistic reasons.

In the mid-sixties, a bad drought resulted in famine because there were no reserve stocks of food. This attracted a number of foreign donor agencies who approached established voluntary organizations to distribute the food aid they brought in. Winin Pereira told one of them that the Mandal was interested in long-term solutions, not merely in immediate relief. The donor agency readily agreed to fund such projects, and the Mandal obtained funds within a few months for small irrigation schemes. "This turned out to be one of the biggest mistakes made by us, because the local collection of funds was stopped and the organization was rapidly expanded. But more important, we were constrained to follow the western path of development because of the donor agency's needs. For instance, they also donated deep-well drilling rigs and sent two volunteers to operate them. Suddenly, there was plenty of money and the Mandal became a big establishment with many workers operating in over a hundred villages."

Yet when the Mandal evaluated its own work, it appeared that the main beneficiaries of mechanization and bore wells were the rich farmers. Nasik is a sugar-cane growing area, and sugar-cane requires a great deal of water. Surface wells provided enough water for small acreages but the bore wells enabled big farmers to irrigate much larger areas. The resulting excessive use of water led to the water table falling and the surface wells going dry. This impoverished many small farmers.

The Mandal closed down its tractor and drilling sections and moved out to areas where the people were poorer – mainly to the Adivasi (tribal) areas. The Adivasis had little land, most of it in hilly, forested areas.

In the early seventies, the Mandal started promoting dairy-farming with cross-breed cows. At that time, they had no reason to doubt all the things they'd heard about cross-breed cows, any more than they had reason to question the conventional wisdom about chemical fertilizers and pesticides. Everyone was promoting these things: agricultural graduates, the government, and above all the aid agencies, importers and manufacturers. At the same time, they were promoting subabul (*Leucaena leucocephala*) as an all-year-round green fodder crop. But subabul contains mimosine, which first made the hair on the tails drop off and could later even kill the animals. Cattle must be limited to 15 per cent of subabul dry matter; only you can't tell a cow to respect such limits. Wildlife was also affected – monkeys and squirrels suffered in the same way. Even worse, the subabul seedlings had been raised in nurseries which were infested with the weed *Parthenium hysterophorus*, which had itself come to India with food grains donated by the U.S. The parthenium spread like wildfire and ruined thousands of hectares of pasture land.

The Mandal also worked in the Thane district of Maharashtra, just to the north of Bombay where many were landless and most farmers owned less than a quarter hectare. At their request, the Mandal gave them interest-free loans to purchase milch buffaloes. The people could earn enough to repay that loan, but never enough to buy a second buffalo when the first went dry. When accounts were kept it was found that, with costs of fodder and concentrate feed so high, the farmers could never be independent. This was even though, as it was later found, some of the buffalo owners gave their dry animals to neighbouring Adivasis – on a payment of Rs 10 per month – to take into the forests for grazing till they were pregnant again, thus passing on part of their losses to the Adivasis.

The farmers also requested loans for chemical fertilizers for their paddy and banana crops. Because of the soil, climate and easy availability of surface water, the area produced some of the best varieties of bananas in Maharashtra. But as the farmers had only small land holdings, it was essential to increase their incomes. "We even tried marketing bananas", Winin explains. "We thought 'Get rid of the middlemen to give the farmers more revenue'. We hired a truck to carry the bananas to Bombay, but the market at Byculla is not really a wholesale market. One of the farmers had to come with the truck, rent a few square metres of market space and sit there till all the bananas were sold. That meant leaving at three o'clock in the morning and often staying all day. It became clear that what they were getting was less than the alleged crooked middlemen were giving them. The middlemen used to cut off a couple of bunches at a time, carry them on their heads to the railway station and sit there until they were sold. We found out that they were serving a proper and useful function." At about that time bananas, which were till then the common person's fruit, began to be exported to the Arabian Gulf countries. The prices rose and the banana farmers benefited a little. But for the others, the local prices, which also rose, made bananas a luxury food which few could afford. Soon what the banana farmers gained in higher prices was lost in general inflation.

"With all these problems, the farmers rarely became self-reliant. The deepest dependency was on the use of commercial inputs, and this was being described as 'progress'."

It was these learning processes that led Winin to stay and work on farming and forestry in a Warli Adivasi area. "It was when working with them that we found that even the most 'appropriate' of western technologies was not merely useless but positively harmful to the people, as well as to the environment. And it was then that we realized that we had more to learn from the Adivasis regarding their traditional knowledge and culture than we had to give.

"The Adivasis used chemical fertilizer because it had been given free to them by the government. After a time, they simply stopped using it or used as little as possible with as much organic manure as they could get; they said chemical fertilizer was ruining the soil. They had also been given seedlings of high-yielding varieties of paddy. They did use them, but rotated them with traditional varieties, to prevent the former from impoverishing the soil. What we learned is that you must also go in for inputs that don't cost a lot of money – organic manures, organic pesticides – and a whole scheme of self-reliant village development, whereby you don't export the richness of the land.

"It is the same with village industry. The best is that which uses the resources of the locality itself to produce things that help local people. The next best is one which uses outside raw materials for things that help people in the village. Third comes those which use local raw materials for things that are sold outside the area. And last are those industries which bring in outside raw materials, process them using village labour, and then export the products. This is what Gandhi said; the truth of it is just as vital now as it ever was."

It was then that Winin saw the need for a holistic understanding of the situation, which led to the formation of a centre for such studies. He stopped all "project" work even though, from the point of view of the foreign donor agencies, the work of the Mandal was running well. It was well-organized, the money was used as intended. It was a "good project". But it was not only the village debtors who were trapped: the donors too were hooked on giving loans. If they stopped, the debtors would stop paying what they owed. It was a mutually buttressing dependency, like the wider world economic system. Donors require clients: and that is another relationship of subordination from which Winin now aims to see the poor emancipate themselves. It is in that context that he seeks to share some of his own experiences in this book.

1. The Other Side of History

Those who now proclaim the virtues of sustainability speak as though industrial society – with its accumulated legacy of liabilities – had not existed for 200 years. The process of colonial expropriation was based on the destruction of sustainable practice. It was a form of development by which the colonial powers exchanged wealth for poverty in the colonies. The methods included the direct transfer of wealth, the de-skilling of people and their induction into the global economy, the dishonouring of traditional knowledge and culture.

The almost universal belief is that we are poor now because of our own deficiencies, and that the West is rich because of their hard work in creating the industrial revolution. In fact, both the poverty and the wealth are mainly the result of colonialism. Further, the process of spoliation did not stop in 1947: it is still going on under such misnomers as "free trade", "technology transfer" and "cultural exchange".

THE TRANSFER OF POVERTY

When Clive first arrived in India, he wrote: "It is a country of inexhaustible riches and one which cannot fail to make its new masters the richest corporation in the world."[1] Britain did indeed become the richest corporation, but mainly by impoverishing the Indians.

With the defeat of the Nawab of Bengal in 1757, Clive claimed from Mir Jafar, the Nawab's successor, £40,000,000 and a personal revenue assignment of £30,000 a year. Although he did not collect the first amount in full, the *Encyclopaedia Britannica* states: "In the context of contemporary values these grants equalled . . . about one-seventeenth of the then annual revenue of Great Britain."

In 1727, the Governor assured the East India Company (EIC, a pioneering transnational company) that every Company servant was allowed the freedom to improve his fortune in any way that he chose.[2] In 1765, the EIC obtained the right to collect land revenue (*diwani*). The profits from this enabled them to increase their armed forces and to monopolize the production and marketing of commodities. At first, no Indian was permitted to buy raw silk until the EIC had bought all its requirements. This "free trade" was then extended to all the EIC's purchases.

Richard Becher, Resident at the Durbar, wrote in May 1769:

> Since the Hon'ble Company have been in possession of the Dewanee the influence has been used in providing . . . such a monopoly, that the chassars, manufacturers . . . have been obliged to sell their commodities at any price . . . If any country merchant . . . attempted to purchase, there was an immediate cry that it interfered with the Company's Investment.

In 1769 the EIC introduced machines to wind silk and reduced the amount of "country wound" silk they bought, so that by about the year 1800, the *chassars* were completely displaced. Coercion was freely used, as Becher showed: "If [the chassars] at any time attempted to break the chain and exercise the rights of a free subject in the disposal of their property, every sphere of tyranny and oppression is exerted over them and their families which perhaps often terminates in immediate beggary and total ruin."

When the EIC found that the *zamindars* (tax collectors) were unable to collect the land revenues the EIC demanded, they appointed as their own agents persons who bid to collect the largest amounts. Becher observed: "What a destructive system is this for the poor inhabitants . . .". If even this did not succeed, the EIC sent in their troops and confiscated lands. Ducarel, the Supervisor of Purnea, reported that: "Two lakh [1 lakh = 100,000] of people perished for want of subsistence; lands are now really lying waste for want of inhabitants, particularly Haveli Purnea, which contained more than 1,000 villages . . .". In spite of these conditions, he said in 1771 "the medium revenue of the province is still supported".

Governor Warren Hastings admitted the use of force to maintain the revenue collection, which meant that whether people were starving or not, the revenue of the EIC remained the same.

The British monopoly in rice meant forcibly seizing it at 15 seers/rupee and reselling at 120 seers/rupee – a huge profit. The result was that people could not buy even a tenth of what was required to keep them from starvation. The Court of Directors observed that "persons of some rank in our service" were responsible for augmenting the distress. They had compelled poor peasants to sell even the seeds required for the next harvest. By this means "one of our writers at Darbar, not esteemed to be worth a 1000 rupees last year, has sent down 60,000 sterling to be remitted home this year".

Peasant revolts were widespread after 1765. Hastings sent a directive to the *zamindars* that if they failed to inform him punctually of the movement of "rebel" forces, their estates would be confiscated. The general result was widespread starvation and misery.

The ill-treatment of workers has been described by Ian J. Kerr: "One railway contractor states: 'The coolie, though fond of money, prefers perfect idleness, and it is frequently necessary to drive him out of his village . . . to force him to earn a good day's wages on the neighbouring railway works.'"[3] Perhaps the "coolie" had other more important work to do, such as growing his own food and looking after his family.

One of the ways in which the British covered up their injustices was to cloak their self-interest with legality. Glaringly unjust laws were passed, but because they had gone through the processes which the British had institutionalized, their implementation became "justice".

In 1835, the EIC introduced its silver rupee as the standard coin in British-occupied India, thereby totally devaluing all Indian currencies. It was said: "No sooner has the small red line which distinguishes the British possession . . . been drawn round a native state, then whatever amount may be fixed as land revenue of the state, that sum in silver rupees . . . must be . . . paid into the government treasury . . .".[4]

The East Indian Railway was bought up by British shareholders in 1879 with the condition that they be paid an annuity of £125 for every £100 share. This guaranteed return covered £60 million of railway investments by the year 1900.

The *Cambridge Economic History of India* (1983) claims that transfers of profits paid to foreign investors, and the salaries and perquisites paid to all Englishmen engaged in commerce here, should not be included in the amount taken out of the country because they were "borne by many other developing countries" also.

One of the items in the transfers was "Home Charges", which included military and civil charges and pensions. These rose from £7 million per annum in the 1870s to over £20 million per year in the 1890s, and much more after that. The British, in effect, made us pay them for oppressing and exploiting us.

From 1858 to 1898 the total amount transferred was more than £1,000 million. From 1898 to 1939, this more than doubled.[5] Add to this what was transferred between 1939 and 1947, and what was taken by way of extortion and illegal exactions, and altogether the amount would probably come to many thousands of millions of pounds. And besides the huge amount extracted in currency, we have to add what was removed in the form of priceless manuscripts, antiques, jewellery, and so on.

All this gives only a faint idea of the extent of our impoverishment during the British occupation. These enormous transfers reduced the funds available for investment here and hence increased unemployment and poverty, and simultaneously financed much of Britain's industrial revolution. Taking all that they transferred and adding interest to it, enormous reparations need to be made before "aid" begins.

TREES: CONSERVATION AND PROHIBITION

Our ancestors had developed, over the ages, a system of renewable resource use that remained environmentally stable till the British legalized the destruction of our forests.[6]

Records show that extensive tracts of India were wooded till the 18th century, because of the importance given to trees in our culture. For instance, the *Agni Purana* (an ancient religious text) states:

> The planting of trees ... (is) conducive to purgation of sin and enjoyment of prosperity ... The supremely wicked man who cuts down trees and thereby stops the passage to wells, ponds and lakes gets his family degraded and even his distant relatives despatched to hell ...

This also shows that the role of trees in the water cycle was realized even at that time.

Some forests were preserved in the form of sacred groves in which felling was severely restricted. Brandis, the first Inspector General of Forests, noted in 1867: "... Sacred groves in India ... are, or rather were, very numerous. ... They ... have been held in veneration by the people from time immemorial. Latterly, however ... many of them have been turned into coffee plantations."

There were other social restraints which served to preserve forests. But the common Indian also clearly understood the ecological reasons for doing so, as the Minutes of Consultation of the Board of Revenue of Madras Presidency of 18 May 1849 show: "The principal Collector of Coimbatore mentions the general opinion among Natives that extensive clearances have an influence on rain". Brandis cited this incident:

> At a coffee plantation ... the proprietor, when preparing the ground which was watered by an excellent spring, was warned by the natives not to clear away trees in the immediate neighbourhood of his spring, but he disregarded their warning, cut down the trees, and lost his stream of water.

Such "superior" knowledge had already resulted in British forests being destroyed. At the end of the 17th century, forests covered one-eighth of England, but by 1825 only a twenty-third, so they turned to India for wood. E. P. Stebbing, the imperial historian, explained the process:

> The first decades of British occupation in India witnessed ... an enhanced rate in destruction of fine timber forests in these regions ... [In 1805] a despatch was received from the Court of Directors enquiring to what extent the King's Navy might in view of the growing deficiency of oak in England, depend on a permanent supply of teak in Malabar ...

In 1799 a timber syndicate, run by a Mr Maconochie from the medical service, cut down over 10,000 teak trees from an area that could yield a harvest of only 1,200 annually. It was to control depredations such as this that a Conservator of Forests was appointed. He was given extraordinary powers in a proclamation of April 1807, which also asserted the Company's "right of sovereignty" over the forests.

W. R. Baillie, in the *Records of the Bengal Government*, stated:

> The proclamation of 1807 ... contained no definition of the term "sovereignty" ... [The Conservator] appropriated to the use of the Company not only the trees of the private forests but even those growing on cultivated lands ... while the proprietor himself, unless expressly permitted by the Conservator, was prevented from cutting a piece of wood on his own property, or removing the young seedling plants that were injuring his land.

By abuse of the word sovereignty, communal forests and every single tree on private holdings became Crown property.

Underwood, a Collector of Malabar, wrote in 1839 that he could not find any record to show that any Conservators had planted trees. No wonder that in 1860 it was noted that "the old forests of Malabar do not now contain much timber". Stebbing said that "in 1800, ... extensive forests of teak flourished throughout the province. These had well nigh disappeared in 1847".

Forests were initially cut down mainly to build the ships with which the British "ruled the waves", since ships built of teak lasted more than 50 years against about ten for those of oak and pine. But in 1853, with the coming of the railways – the backbone of colonial rule – the need for wood for sleepers and locomotive-fuel increased rapidly. However, it was the 1857 War of Independence that sent the British into a frenzy of railway construction. Stebbing said:

> The incidence of the mutiny at once threw into glaring relief the paucity of the communications in the country. The necessity for railway construction if only to facilitate the movement of troops and their equipment had become vital. The Government set themselves

feverishly to work ... The urgent demands for timber to provide the sleepers for the new railway lines were met in the time-honoured fashion, and great forest areas ... which ... had hitherto remained untouched by man, were ruined in order to supply the demands.

It was not only the forests which were destroyed to build the railways. Harappan bricks from the important Indus Valley prehistoric site, thousands of years old, were used to provide ballast for about 100 miles of track nearby.[7]

Realizing that the forests would not last much longer, the British decided to establish monopoly rights in the (British) public interest. In 1865 the first Forest Act was passed, with the preservation of forests as its ostensible purpose.

The Act gave the Governor-General of India extensive powers over all forested lands, without, it was explicitly stated, contradicting "any existing rights of individuals or communities". But, in prohibiting access to forests and all customary economic and social activity in the forests, this is exactly what the Act did.

C. C. Wilson, Chief Conservator of Forests, described its consequences:

> The dwellers in the countless villages all over the country had, from time immemorial, obtained a great part of their daily needs from the jungles. First and foremost was the question of fuel with which to cook their food. Without that they could not live. Then there were small timbers for building without which they would have no shelter, ploughs without which they could not cultivate the ground, grazing without which their cattle would die, green-leaf manure for their fields, tanning bark for their leather, bamboos for a dozen different purposes. And these were vital to their well being ... And then an authority came into being which denied them what they had always looked upon as their rights. They fought most bitterly and indeed understandably, against the new tyranny.

Baden-Powell drafted a new act in 1874 which effectively destroyed the rights of the people. He did this by the simple process of saying that they did not really have rights, but only privileges. And privileges, according to him, could be abrogated at any time by the State. Occasionally, even the flimsy cloak of legitimacy was cast aside. One C.F. Amery observed with unusual candour: "The right of conquest is the strongest of all rights – it is a right against which there is no appeal."

An "improved" act in 1878 prescribed punishment of six months imprisonment or a fine of Rs 500 for any person who "hunts, shoots, fishes, poisons water or sets traps or snares", in addition to what was prohibited earlier. And Baden-Powell insisted that for enforcement of the Act "arrest without warrant is absolutely essential".

The Madras Government insisted that: "the forests are, and always have been common property". The Governor recorded:

> This Bill if passed, will . . . give rise to grave dissatisfaction . . . It will do so, not because the native people are averse to the maintenance and protection of forests, but because the basis and principle of the Bill is the ultimate extinction of all private rights in or over forest or waste land and their absorption by Government. As regards interests of the villages . . . it is a *Bill for confiscation*, [author's emphasis] instead of protection . . . The powers proposed to be given to the Police are arbitrary . . . viz, arrest without warrant or order by any Policeman of any person suspected of having been concerned at some unknown time (2 years previously) of being concerned in a forest offence (taking some wild bee's honey from a tree or the skin of any dead animal) . . .

Brandis evaded the issue by totally denying the very existence of communal forests: "I have not been able to discover any facts indicating the existence of communal Forests". And with this convenient non-discovery the Act was passed.

The average annual forest revenue for the whole of India was Rs 3.74 million in 1864–9, increasing to Rs 30.19 million in 1937 and Rs 124.37 million in 1944.

A Forest Institute did exhaustive research on "spars for aeroplanes, poles for gun carriages, stocks for army rifles", all "in the service of India". Hundreds of thousands of tons of timber and fodder grass were exported to help war operations in 1914–18; while further thousands of hectares of forest were denuded during World War II.

While the people were deprived of forest lands under the excuse of protecting them, vast areas were given to the British to develop. A writer in 1913 describes the process:

In Assam, . . . special grants of waste land in large blocks were made on easy terms from 1860 onwards; and there are now 200,000 acres under tea in Assam proper, and about as much again in Sylhet and Cachar, representing the enterprise and outlay of English companies. Again, in the Basti and Gorakhpur districts of the United Provinces where there was much unreclaimed waste land, some 600,000 acres were given out in large estates during the thirties. The object was to attract capital and enterprise to the task of reclamation. . . . On one estate the English zemindar constructed 185 miles of canals, on another a protective embankment three miles in length, while in another wells, and tanks, and villages were created.[8]

There were other consequences of the policies that robbed India of its forest wealth. For instance, a tax was levied on firewood which increased the cost of living and made most forest-based industry uneconomic. Brandis said:

> It is in accordance with immemorial usage to permit villagers to cut firewood free . . . Industries such as iron-smelting, sugar-boiling and the manufacture of Indigo should certainly by all means be encouraged, but this must be done by promoting the growth of wood, and by discouraging its waste.

This did not prevent the government from permitting the British Aska Sugar factory in Ganjam to use an unlimited quantity of firewood at a fee of Re 1 per year. Their annual requirement was about 10,000 tons. On the other hand, the poorest had to pay Re 1 per ton. The industries which they claim to have encouraged were in fact ruined.

The destruction of agriculture is described by W. Wedderburn:

> Under our land administration no care has been taken to preserve these communal forests or to help the peasants . . . The cultivators are thus stinted both in grass and wood, and are driven to use as fuel the droppings of cattle which are their only manure . . . The plough cattle are decimated each hot season in villages deprived of their communal preserves of wood and grass . . .

Animal dung was not the only fertilizer loss. Traditional farming used large quantities of dried leaves and twigs for manure. This tremendous loss of soil nutrients caused an equally enormous reduction in food production and has resulted in long-term deterioration of the soil.

The people most affected by the Acts were the Adivasis who depended totally on the forests for their livelihood. They naturally revolted against the British and never gave up struggling for their rights. Therefore the British, so concerned with human rights now, declared them to be "criminal tribes".

These Acts were the beginning of the long slide to unsustainability.

Today, the government still claims proprietory rights over the forests. The proposed new Forest Bill takes no account of traditional forest usage and completely disregards the basic needs of villagers. Small artisans who need bamboo and other forest products are denied these so that the insatiable appetites of paper mills can be fed.

The two major instruments devised for reversing the trend of deforestation – the Forest Conservation Act (1980) and the National Wasteland Development Board – are in violent conflict. The Forest Conservation Act took away the power of the State Governments to dereserve a forest area, on an assumption that the Central Government would be able to view deforestation proposals more objectively than a State Government.[9]

The 1980 Act has not deterred the central government from permitting the clearance of thousands of hectares of forest land in order to accommodate factories, dams and mining projects. The clearance of Sardar Sarovar (Narmada) in less than a day by the Prime Minister, under pressure from the Gujarat politicians – when the case had been pending for years with the Government – is the most shocking example of the absolute irrelevance of the Forest Conservation Act in the preservation of forests.

The conflict between this Act and the activities of the National Wasteland Development Board is illustrated by the inability of voluntary agencies to obtain any land for afforestation programmes, because Central Government clearance has to be obtained. Such clearance has never been given. Under Section Two of the Act the diversion of forest land for non-forest purposes is prohibited. If the Government is really serious about participative afforestation of wastelands, non-officials must be involved, and the stranglehold of the Forest Conservation Act broken.

AGRICULTURE: FROM FEAST TO FAMINE

Harappan sites show that barley and wheat were cultivated and ground to flour, and that cotton was planted, nearly 5,000 years ago. Flint tools and sickles were also widely used at that time.

Just before the Europeans came to India, agriculture here was as advanced as – if not superior to – European agriculture. There was a large variety of appropriate implements in use, the agricultural practices were as good as any used now. Extensive famines, as those which later happened under the colonial regime, did not occur.[10]

Alexander Walker, resident at Baroda, wrote (around 1820):

> The Hindoos have been long in possession of one of the most beautiful and useful inventions in agriculture. This is the Drill Plough. This instrument has been in use from the remotest times in India. They have different kinds of ploughs . . . adapted to different sorts of seed and soils. They have a variety of implements for husbandry purposes, some of which have only been introduced into England in the course of our recent improvements. The numerous ploughings of the Hindoo Husbandman have been urged as a proof of the imperfection of his instrument; but in reality they are a proof of the perfection of his art. It is not only to extirpate weeds that [he] reploughs and cross-ploughs; it is also to loosen the soil, apt to become hard and dry under a tropical sun.

He gives examples of the growing of green fodder throughout the year, of intercropping, crop rotation, fallowing, manuring and so on. There was also a deep understanding of the nature and quality of soils and all the processes involved in producing a wide variety of crops. He adds: "There are more kinds of grain cultivated perhaps than in any other part of the world . . . I am at a loss to know what essential present we can make to India". Walker is similarly impressed with Indian techniques of irrigation, commenting on "the vast and numerous tanks, reservoirs and artificial lakes as well as dams of solid masonry" which he had seen constructed.

An agricultural scientist, Albert Howard, who held many posts in India in this century, wrote that Indian farmers used compost and organic manures which ensured that they could continue farming on the same land for more than 2,000 years without a drop in yields. He adds that the crops "were remarkably free from pests".[11]

One of the important reasons for the decline of this excellent farming system was the land revenue extracted by the British which was theoretically fixed at 50 per cent, but in practice was much higher. In one area, the taxes amounted to over 63 per cent of the net profit of the farmer in a good year. If, for any reason, he had a bad crop he would almost surely make a loss because the amount of tax remained fixed. Boswell, a Collector of the district, said: "The idea is to get as much as possible out of the ryot [peasant farmer] and when no more is to be got out of him, let him go to the devil!"[12] This alone resulted in about a third of the irrigated land going out of cultivation there.

Albert Howard maintains that: "It is a noteworthy fact that the excessive development of alkalis in India is the result of irrigation practices modern in their origin and modes and instituted by people lacking in the traditions of the ancient irrigators, who had worked these same lands thousands of years before."

The traditional irrigation system was based on village cooperation and was supported by local rulers, but when revenue was extracted and transferred for Home Charges, the people could no longer take care of their systems and they gradually fell into disuse. Even where the British extended irrigation, it was not to help the farmers or grow more food; they promoted the cultivation of sugar cane because its advantage "over any other crop is the great amount of water it takes in a restricted area, giving a very big revenue return".[13]

The forcible conversion of food land to cash crops resulted in a further decline in food output. During the US Civil War, the British compelled Indian farmers to grow cotton instead of food. But when the war ended, the cotton market collapsed. Millions of peasants were unable to convert back into food production with the result that "from 1865 through 1900 India experienced the most severe series of protracted famines in its entire history."[14]

The aggregate result of British policies is evident from the fact that there was practically no growth in food output during the last decades of British rule, but that growth increased rapidly after independence.

. But the pattern set then has intensified in the years since independence. Unsuitable agricultural techniques such as the use of synthetic fertilizers and pesticides are continuing to destroy stable traditional ecosystems, and the use of a few high-yielding crop varieties has resulted in the elimination of thousands of traditional varieties, with the ensuing loss of genetic resources and the increasing danger of large scale famines.

EDUCATION: LEARNING TO BE BRITISH

Mahatma Gandhi, in a 1931 address to the Royal Institute of International Affairs, London, stated:

> I say without fear of my figures being challenged successfully that today India is more illiterate that it was fifty or a hundred years ago . . . Because the British administrators when they came to India, instead of taking hold of things as they were, began to root them out. They scratched the soil . . . and left the root like that and the beautiful tree perished.[15]

The British rushed to prove Gandhi wrong but surveys of education in India (carried out earlier by the British themselves) showed not only that he was right, but that literacy was higher in India than in England in those days. These surveys were carried out in the Madras Presidency in 1822 by Governor Munro, in the Bombay Presidency in 1824, in Bengal-Bihar in 1835 by William Adam, and in the Punjab in 1882 by G.W. Leitner, Director of Public Instruction.

Adam observed in 1835 that there was a school in nearly every village in Bengal-Bihar. Governor Munro found the same in Madras. Prendergast, of the Governor's Council in Bombay, stated in 1821:

> there is hardly a village, great or small, throughout our territories, in which there is not at least one school, and in larger villages more; many in every town . . . where young natives are taught reading, writing and arithmetic upon a system so economical . . . and at the same time so simple and effectual that there is hardly a cultivator or petty dealer who is not competent to keep his own accounts with a degree of accuracy . . . beyond what we meet with amongst the lower orders in our own

country; whilst the more splendid dealers and bankers keep their books with a degree of ease, conciseness and clearness, I rather think fully equal to those of any British merchants.

The population of England in 1811 was about 75 per cent that of Madras Presidency. However, the number of those attending charity, Sunday, circulating, and other schools in England was about 75,000 whereas at least twice that number attended schools in Madras and many others in Madras were taught at home by private tutors. Further, more than half of those 75,000 attended Sunday school for less than three hours a week, with the average time spent in school being about one year in 1835 and two years in 1851.

Adam said that by the end of schooling in Bengal-Bihar the student was expected to read a number of books as well as be qualified in accounts (agricultural as well as commercial) and the writing of letters and petitions.

Walker, in his note on education in the Malabar area (about 1820) stated:

> The children are instructed without violence and by a process peculiarly simple. The system was borrowed from the Bramans and brought from India to Europe. It has been made the foundation of National Schools in every enlightened country. Some gratitude is due to a people from whom we have learnt to diffuse among the lower ranks of society instruction by one of the most unerring and economical methods which has ever been invented. The pupils are the monitors of each other . . .

This monitorial method was taken to Britain by Bell, a chaplain at Madras, and came to be called the "Bell and Lancaster" system.

Higher education was also widespread and covered a number of subjects, not just the Vedas, as commonly thought. Adam notes that all castes could join these colleges but the study of law, philosophy and sacred poems was limited to Brahmins.

A detailed account of the University of Nuddea, published in 1791, stated:

> The grandeur of the foundation of Nuddea University is generally acknowledged. It consists of 3 colleges . . . each of which is endowed with lands for maintaining masters in every science. Whenever the revenues of these lands prove too scanty for the support of the Pundits

and their Scholars, the Rajah's treasury supplies the deficiency . . . In the College of Nuddea alone, there are at present 1100 students, and 150 masters . . . In Rajah Roodre's time [about 1680] there were no less than 4,000 students and masters in proportion . . . Their method of teaching is this: two of the masters commence a dialogue . . . on the particular topic they mean to explain. When a student hears anything . . . that he does not perfectly understand, he has the privilege of interrogating the master about it. They give the young man every encouragement to communicate their doubts, by their . . . patience in solving them.

This system was maintained by grants of rent-free land given to teachers, by which they supported themselves and their pupils. The British deliberately destroyed the system merely by taking possession of the lands. According to Leitner: ". . . throughout the country, by far the large majority of schools held on the grant of rent-free system were resumed".

The indigenous system was also supported by local taxes. The *Public Despatch* of Bengal of 3 June 1814 stated: "We refer . . . to that distinguished feature of internal polity which prevails . . . by which the instruction of the people is provided for by a certain charge upon the produce of the soil and other endowments in favour of the village teachers". Rulers also made assignments to scholars, poets, medical practitioners and others. In the Madras Presidency "as late as 1801 over 35 per cent of the total cultivated land . . . came under the category of revenue-free assignments". Munro reduced this to a mere five per cent. This occurred in other districts too.

Finally, it was Macaulay who completed the destruction of Indian education with his orders to produce "a class of persons, Indian in blood and colour, but English in taste . . .".[16]

One of the reasons why the literacy rate dropped after the British introduced their system was that theirs was much more expensive to run. Leitner calculated the cost of a Government-aided school in 1881 to be 15 times higher than that of an indigenous one. Apart from that, the new system was irrelevant to the people. Leitner cites a report of 1882–83: "If it were not for the pressure put on by the tahsildars [a higher caste] sub rosa, there will be few agriculturists' sons in the Government schools at all. The reason is that the natives say that a boy is lost to them when he enters a Government school . . . he is of no use . . .".

A. D. Campbell, a Collector of Bellary, gave further reasons for the decline in education in a report of 1823:

> I am sorry to state that this is ascribable to the gradual but general impoverishment of the country. The means of the manufacturing classes have been greatly diminished by the introduction of our own European manufactures, in lieu of the Indian cotton fabrics . . . The transfer of the capital of the country, from the native governments and their Officers who liberally expended it in India, to Europeans, restricted by law from employing it even temporarily in India and daily draining from the land has likewise tended to this effect . . . The greater part of the middling and lower classes of people are now unable to defray the expenses incident upon the education of their offspring, while their necessities require the assistance of the children as soon as their tender limbs are capable of the smallest labour . . . Of the 533 institutions for education now existing in this district, I am ashamed to say, not one now derives any support from the state.

The Colleges were destroyed in the same way as the schools: by withdrawing funds. In Pune, the Peshwas distributed several *lakhs* every year to learned Brahmins. By 1824, this was reduced by the British to about Rs 35,000.

The result of all this was a decline in the percentage of boys attending school as against the number of boys of school going age. This was 33.3 per cent in 1822 but with the introduction of the British system this dropped to 12.58 per cent in 1879–80. The figure rose to 27.8 per cent only in 1899–1900.

SCIENCE AND TECHNOLOGY: INDUSTRIAL ESPIONAGE

Our scientific and technological history began thousands of years ago with accurate astronomical observations and the standard dimensioned bricks of the Harappans.[17]

The British came to India not only to trade but also to learn, as the mathematician Reuben Burrow showed in 1783:

Notwithstanding the prejudices of the Europeans of last century in favour of their own abilities, some of the first members of the Royal Society were sufficiently enlightened to consider the East Indies . . . as new worlds of science that yet remained undiscovered. They . . . seemed extremely desirous of possessing the literary treasures of these unexplored regions of knowledge of which they had formed such sanguine expectations.

Dr H. Scott was requested in 1788 by the President of the Royal Society to send the latter information on Indian technology. He wrote in reply: "The people of this country . . . have retained . . . perhaps, from ages very remote a considerable degree of civilisation. I often think that their arts improved by the practice of so many years might afford matter of entertainment and instruction to the most enlightened philosopher of Europe."

Burrow, Scott and many others sent a considerable amount of information to Europe, much of which was copied or incorporated into European technology.

Astronomy

One of the most astonishing things about Indian astronomy is that accurate observations appear to have been made more than 6,000 years ago. These observations resulted in advanced theories of the solar system. John Playfair, FRS, Professor of Mathematics at the University of Edinburgh, proved the antiquity, originality and accuracy of Indian astronomy in a paper, *Remarks on the Astronomy of the Brahmins*, written in 1790.

European scientists found that Indian astronomers used a number of tables which were remarkable in that they went back to the year 3102 BC, the beginning of the era known as Kaliyug. Playfair proved that the tables could not have been calculated back from recent data but were derived from "observations made in India, when all Europe was barbarous".

The various tables collected showed differences in calculations which meant that they were not the creations of just one individual genius, but were the result of a number of intelligent people using different approaches. He said: "The methods of this astronomy are as much diversified as we can suppose the same system to be, by passing through the hands of a succession of ingenious men, fertile in resources, and acquainted with the variety and extent of the science which they cultivated".

The superiority of Indian astronomy over that of the Europeans at that time is confirmed by the fact that Aryabhata (about 500 AD) knew that the earth was round and that it rotated on its axis. He calculated the length of the year as 365.3586 days. The third and fourth satellites of Jupiter and the sixth and seventh of Saturn were known to ancient Indian astronomers: in Europe, the former were unknown before 1609 and the latter were only observed in Europe by Herschel in 1789.

Mathematics: relations of inequality

India is well known for giving the world the zero and the decimal system. What is not so well known is that we were well advanced in other aspects of arithmetic, algebra and geometry.

Indian mathematicians used very fast mechanical methods for multiplication. M.A. Williams states in his *Annals of the History of Computing* that one ancestor of the modern computer was John Napier, of logarithmic fame. He made a device to simplify multiplication, using a set of rods called Napier's Bones. Williams claims that the inspiration for Napier's Bones "was surely an ancient method of multiplication . . . [which] found its way from India to Italy in the 14th century."[18]

Playfair stated that Indian astronomers, in calculating the occurrences of eclipses, used a simplification of a general formula which showed that they understood the theory of errors.

He showed that they used a value of Pi that was much more accurate than that calculated by the Greeks, and equal in accuracy to that in use in Europe in his time. The derivation of this value required "an operation which cannot be arithmetically performed without the knowledge of some very curious properties of [the circle] and, at least, nine extractions of the square root, each as far as ten places of decimals".

The discovery of the Binomial Theorem was credited to Newton but two British mathematicians, Reuben Burrow and H. T. Colebrook, showed that it was in use by Indians much earlier.

Colebrook showed "that the Hindus were in possession of algebra before it was known to the Arabians", and goes on to mention three major instances where Indians had anticipated recent European discoveries. Bhaskara had proved algebraically Pythagoras' theorem centuries before Wallis rediscovered the proof; the general solution of indeterminate problems of the first degree was known to Bramhagupta and only rediscovered by Bachet de Mezirac in 1624; and a general method for the solution of indeterminate problems of the second degree was given by Bramhagupta, "a discovery, which, among the moderns, was reserved for Euler in the middle of the last century".

Medicine

The Ayurvedic system of medicine has existed for many centuries. The active ingredients in many of the herbs used are now being identified, clinically tested, and found to be highly effective.

A very old and successful Indian practice here was inoculation for smallpox. A detailed account of this was given by J.Z. Holwell, FRS, in a paper to the College of Physicians, London, in 1767. Holwell said that his reason for writing the paper was "the great benefit that may arise to mankind from a knowledge of a foreign method".

He wrote:

> Inoculation is performed . . . by a particular tribe of Bramins, who are delegated annually for this service from the different Colleges of Bindoobund, Eleabas, Banaras, &c over all the distant provinces . . . [The physician] with a small instrument, wounds, by many slight touches . . . just making the smallest appearance of blood . . . Matter from inoculated pustules of the preceding year [is placed on these scratches], for they never inoculate with fresh matter, nor with matter from disease caught in the natural way . . . When the before recited treatment of the inoculated is strictly followed, it is next to a miracle to hear, that one in a million . . . miscarries under it.

Inoculation was introduced into Britain only after the wife of the ambassador to Turkey publicized it in 1720. Jenner developed his vaccination method only in 1796, and the British then banned the

Indian system in order to impose Jenner's on us. This resulted in widespread smallpox epidemics in the 19th and 20th centuries.

Physicians also believed that diseases were caused by what we now call bacteria. Holwell stated: "They lay it down as a principle, that ... the acting cause ... is multitudes of imperceptible animalculae floating in the atmosphere; that these are the cause of all epidemical diseases ... [and] that these bodies are imperceptible ... to the human organs of vision".

Surgery for tumours, ulcers, cataract, kidney stones and other purposes was practiced from ancient times. A Colonel Kyd stated that plastic surgery was quite common, and he sent to London a quantity of "cement" used "for uniting animal parts".

The destruction of the colleges and the general impoverishment of the people led to a decline in the practice of indigenous medicine which caused increased illness and mortality.

Steel

The Damascus swords used by the Saracens were famous, but what is not so well known is that they were made of Indian steel called *wootz*.[19]

Damascus blades were in use as far back as the time of Alexander the Macedonian (about 325 BC); *wootz* was produced much earlier than that. There was nothing anywhere else in the world to equal this steel, which made possible the production of blades which were exceptionally strong and flexible.

A variety of irons and steels were produced in furnaces in which iron ore was mixed with charcoal. "Depending on the amount of charcoal in the mixture, the product was a wrought iron, which has a very low carbon content, or a pig iron containing more than 4 per cent carbon. The Indian steelmakers manufactured *wootz* either by adding carbon to wrought iron or by removing carbon from pig iron." This clearly shows that they understood the role that carbon played in determining the properties of steel. However, it was only in 1821 that the Parisian, Breant, found out that the properties of *wootz* were due to its high carbon content.

The production of *wootz* was quite complicated and involved heating and cooling at several specific rates, with temperatures ranging up to 1,200 degrees centigrade.

A sample of *wootz*, sent by Dr H. Scott to the Royal Society in about 1790, was analysed by several experts. One user stated: "If a better steel is offered to me, I will gladly attend to it; but the steel of India is decidedly the best I have yet met with."[17] Scott wrote that *wootz* "is employed . . . for cutting iron on a lathe, for chisels for cutting stones, for files and saws and for every purpose where excessive hardness is necessary". The surest proof that hard chisels were in use in India for many centuries is given by the evidence of huge, ancient caves hollowed out of solid rock, chip by tiny chip, and decorated with intricate, stone sculptures.

J. Campbell, Assistant Surveyor General of the Madras Establishment, wrote in 1842 that the methods of manufacture of iron in India were more efficient and cheaper than those current in England. "From what I have seen of Indian iron, I consider the worst I have ever seen to be as good as the best English iron . . ."

The Forest Acts prevented the smelters from collecting free wood of the specific trees used for making charcoal. Brandis mentioned that a Mr Bruce Forte in 1864 "correctly states that it is owing most likely to the greatly increased price of charcoal that the number of native iron-smelting furnaces in the Salem District has decreased of late years."[6]

The industry was also destroyed by the conscription of its workers. Dr Heyne, a botanist at Madras, said that the iron-makers lived under "constant dread lest they should be pressed for the purpose of carrying burdens for strangers [i.e. the British] from one village to another, a thing which often happens in the very season when it is in their power to employ their time to most advantage to themselves". Heyne also mentioned that "before the famine [of 1790], there were besides 40 smelting furnaces, a great number of silver and copper smiths, here, who were in a state of affluence; but their survivors are now poor and in a wretched situation".

Finally, it was the dumping of British iron that completed the destruction of the industry. Campbell stated: "Among the most extensive of the exports of England to India, is the trade of bar iron, which to Madras alone amounts to 1000 tons per annum; . . . India is known to produce malleable iron of a superior quality".

Besides *wootz*, the Iron Pillar, near the Kutub Minar, shows the high level of technology achieved. This pillar, cast as a single unit of many tons, has not rusted for centuries. It was classed as "one of the most marvellous items of antiquity in the world".

Efforts are now being made by the PPST Foundation to collect information from the few still living who remember the processes for producing *wootz*.[20] Many more technologies need to be recovered in a similar manner.

Textiles

Cotton was grown in India by the Harappans more than 5,000 years ago, and textiles have been made here ever since.

Textiles provided employment for hundreds of thousands of handloom weavers in the country and much was exported all over the world.[21] The Abbé de Guyon described the situation in the middle of the 18th century: "People of all nations, and all kinds of mercantile goods . . . are to be found at Ahmedabad. Brocades of gold and silver, carpets with flowers of gold . . . satins, and calicoes, are all manufactured here".

The famous calicoes ranged from the finest muslins to the cheaper varieties of coarse cloth. There were calicoes patterned on the loom, woven from different coloured threads, and others dyed after weaving. The colours remained bright after washing, and this proved to be their great attraction. The best were painted rather than printed, the dyes and mordants being applied to the cloth, not with a wood-block but free-hand with a brush.

The woollen industry in Kashmir produced extraordinary cashmere shawls, silks were produced in western India, while some fabrics were made with cotton and silk mixed.

A 19th century manuscript by G. Roques, which presented a detailed account of the textile manufacturing processes, proved beyond doubt that mordants for dyeing were used here earlier than in Europe. India's influence was, in fact, crucial to the European textile industry. Imported printed fabrics were so expensive there that only the nobility could afford them. Local imitations were at first banned but, because they were so popular, manufacture was permitted from the 1750s.

In 1676, in England, William Sherwin obtained "a grant for 14 years of the invention of a new and speedy way for producing broad calico, which being the only true way of the East India printing and staynning such kind of goods".

The destruction of this magnificent indigenous industry was

begun by the EIC. In *Consideration on Indian Affairs* (1772), W. Bolts described the process in harrowing detail:

> With every species of monopoly therefore, every kind of oppression to manufacturers, of all denominations throughout the whole country, has daily increased; in so much that weavers, for daring to sell their goods . . . have, by the Company's agents been frequently seized and imprisoned, confined in irons, fined considerable sums of money, flogged, and deprived, in most ignominious manner, of what they esteem most valuable, their crafts. Weavers also, upon their inability to perform such agreements as have been forced from them by the Company's agents . . . have had their goods seized, and sold on the spot . . . and the winders of raw silk . . . have been treated also with such injustice, that instances have been known of their cutting off their thumbs, to prevent their being forced to wind silk.[22]

From 1770 to 1789 the Surat Murshirabad silk trade dropped from 20,000 to 1000 maunds (1 maund equals 40 kilograms) per year. After 1879 the silk trade of the Indian merchants died out completely. The British later shipped raw cotton to Lancashire and sent finished cloth back to India.

J. Campbell stated in 1842:

> In the commerce between India and England, a source of deep injury to the former country arises from England having deprived her of the trade in cotton cloth, the manufacture of which was, but a few years ago, one of the most valuable and extensive of Indian products; while from no other having been as yet introduced as an export to balance the imports from England, it has become necessary to drain India of her specie to pay the expenses of the Government . . ."[23]

As long as they could manipulate it and benefit from it, the British insisted on free trade. Now that Indian textiles are again competing on the world markets, free trade is rapidly being forgotten, and Indian exports are being controlled by the use of the GSP and GATT and other trade barriers.

Yet our present textile policy is designed to help big manufacturers using imported synthetic fibres, in the expectation that exports will increase to cover the costs of the liberal import of luxuries. The result will be that the handloom natural fibre industry

– which employs hundreds of thousands – will again be driven into oblivion.

THE INDUSTRIAL INVERSION

This has been only a brief account of our ancient science and technology. As for literature, music and art, it may be noted that our great epic poets were composing 1,200 years before Chaucer, that our tablas are the only drums that have harmonic overtones and that a single cave in Ellora – the Kailash temple – is larger than the Parthenon and required the removal of 85,000 cubic metres of stone.

Our technology as it existed before the British arrived here was highly appropriate. The steel, dye, textile and other industries were all situated near sources of raw materials. Many of them were conducted during slack agricultural periods, complementing farming work and providing full employment to the people. It was these factors that made them efficient.

Dharampal writes:

> Instead of being crude the processes and tools of 18th century India appear to have developed from a great deal of sophistication in theory and an acute sense of the aesthetic . . . In the context of the values and aptitudes of Indian cultural and social norms and the consequent political structure and institutions, the sciences and technologies of India, instead of being in a state of atrophy, were in actuality usefully performing the tasks desired by Indian society . . . In most respects the political and social ideas of India and its legal and administrative arrangements as well as sciences and technologies had achieved maturity and balance . . . Its social and political structure at this period . . . was able to provide basically a similar sort of freedom, well-being and social security as is at present available in much of the European world."[23]

The British claim that one of the biggest benefits to India of their colonial rule was modern industrialization. If this was so, it should be evident in the growth of national income. However, the per capita increase in this from 1860 to 1900 was 0.42 per cent annually compounded. And this increase came after a considerable earlier decline caused by the destruction of our local industries and with a population kept firmly "planned" by widespread famines at a growth rate below 0.4 per cent per year.

Today, the destruction of whatever little self-reliance we have left is continuing with the free import of western "high" technology and goods, and the exports of our raw materials. Village artisans continue to be under relentless pressure from mass production industry encouraged under the drive for efficiency, even though the latter is increasing unemployment.

NOTES AND REFERENCES

1. "Introducing World Development", *New Internationalist*, December 1974.
2. This account of the East India Company's activities is taken from Suranjan Chatterjee, "New Reflections On The Sannyasi, Fakir And Peasants War", *Economic and Political Weekly*, 28 January 1984.
3. Ian J. Kerr, "Working Class Protest in 19th Century India", *Economic and Political Weekly*, 26 January 1985.
4. Arun Banerji, "British Rule and the Indian Economy", *Economic and Political Weekly*, 1984, annual number.
5. A.K. Banerji, "Aspects of Indo-British Economic Relations 1858–1898", 1982, quoted in A. K. Bagchi, "British Imperialist Exploitation of India", *Economic and Political Weekly*, 5 March 1983.
6. This account is adapted from "Social Restraints on Resource Utilization: The Indian Experience", and "The Story of 'Scientific Forestry' in India", *PPST Bulletin*, May 1983.
7. Ed: G. L. Possehl (ed), *"Harappan Civilization: A Contemporary Perspective"*, Humanities Press, quoted in *Scientific American*, August 1984.
8. G. F. Keatinge, "Agricultural Progress in Western India", *Poona Agricultural College Magazine*, July 1913.
9. M. N. Buch, "Afforestation Utterances, Deforestation Policies", *Sanctuary*, September 1989.
10. Much of the following has been adapted from Dharampal, *Indian Science & Technology in the Eighteenth Century*, (Hyderabad: Academy of Gandhian Studies, 1971). See also David Arnold, *Famine, Social Crisis and Historical Change*, (Oxford: Blackwell, 1988).
11. Albert Howard, *The Soil & Health: A Study of Organic Agriculture*, (New York: Schooken Books)

12. G. N. Rao, "Transition From Subsistence to Commercialised Agriculture: A Study of Krishna District of Andhra, c. 1850–1900", *Economic and Political Weekly*, 29 June 1985.
13. I. Chitelen, "Origins of Co-operative Sugar Industry in Maharashtra", *Economic and Political Weekly*, 6 April 1985.
14. *Encyclopaedia Britannica*, 15th Edition
15. This discussion is mainly based on Dharampal, *The Beautiful Tree: Indigenous Indian Education in the Eighteenth Century*, (New Delhi: Biblia Impex, 1983).
16. Macaulay, "Minute on Indian Education", quoted in Edgar Faure et al., *Learning to Be*, (Paris: UNESCO, 1972).
17. The discussion of Science and Technology is mainly adapted from Dharampal, *Indian Science and Technology in the 18th Century*, (Hyderabad: Academy of Gandhian Studies, 1971).
18. "Dem Bones", *Scientific American*, May 1984.
19. This account is adapted from Oleg D. Sherby, and Jeffrey Wadsworth, "Damascus Steels", *Scientific American*, February 1985.
20. PPST: Patriotic and People-oriented Science & Technology Foundation, Madras.
21. This account is mainly based on Claude Alvares, *Homo Faber: Technology and Culture in India, China and the West 1500 to the Present Day*, (Bombay: Allied Publishers, 1979).
22. From Chatterjee, op. cit. (note 2).
23. From Dharampal, op. cit. (note 17).

2. The Sustainable Lifestyle of the Warlis

Adivasis are considered the original inhabitants of India. They were – and often still are – believed to be "primitive tribals" by the "civilized mainstream", a classification which has rationalized the oppression of the Adivasis. However, in spite of the invading "civilization", they still live in a manner which has sustained them for millennia. Nature in the form of tropical forests is bountiful, but Adivasis use its products sparingly.

There are several clans of Adivasis occupying part of the region north of Bombay, the most numerous being Warlis, who call themselves the Kings of the Jungle. Though references are made mainly to Warlis, what follows applies to other Adivasi clans as well.

JUSTIFYING OPPRESSION

In the year 1876, W. R. Pratt, the British Collector of Surat, wrote: "The cultivators from whom a mass of revenue is derived are little higher than monkeys in the gradation list of animated nature."[1]

Those who dominate have the advantage of being able to impose definitions. To civilize, according to European colonizers, was to "bring out of a state of barbarism", to "instruct in the arts of life", to "enlighten and refine", to "elevate the rude, unpolished world". Those who defined these comparative terms claimed to possess all the attributes which made them civilized. The rest – who differed from the European way of life – became, by definition, barbarians. These definitions still serve the objectives of those who wish to dominate.

The colonizers found it necessary to practice such mental subjugation to justify their physical oppression. The then new "science" of anthropology, which produced the "gradation list", attempted to rationalize this need. The "others" were "proved" to be irrational, unpolished, ignorant, unscientific, lazy – the opposite of what the colonizers imagined themselves to be.

Later, those who accepted British rule and its values were considered civilizable. The Adivasis, however, were among the few who did not give up their struggle against the invaders, so they were labelled primitive tribes – some even criminal tribes – because of their violent resistance to domination.

But the ability of the Adivasis to supply "a mass of revenue" effectively positioned poor Pratt and his tribe well below the former in the gradation list.

To strengthen their claims against such arguments, the old definitions were changed to more "scientific" ones. "Primitive" societies are now defined as those which are undifferentiated as opposed to "advanced" ones which are differentiated. A society is claimed to advance according to the degree to which it develops specialized economic and social institutions. Such a definition is also arbitrary and self-serving.

Civilization cannot simply consist of *haute cuisine* and *haute couture*, Wagner and Van Gogh, monuments, moon walks and military might. Common sense tells us that the basic requirement for a society to claim such a title is that it should promote justice and preserve its own life-support system. If a highly differentiated society is self-destructive, or survives on the exploitation of others, it forfeits its claim to be civilized.

The ability to preserve the integrity of the whole of creation – a necessity for justice and sustainability – then becomes a measure of civilization. Differentiation is irrelevant and may even be incompatible with sustainability.

THE PROGRESS OF OPPRESSION

Jambu Bhoye, an Adivasi from Jadhavpada, recalls his grandfather telling him that Adivasis occupied vast lands in common before the British arrived. On these they practised shifting cultivation, but used only as much land as was needed to supply their annual requirements. Shifting cultivation is being blamed for much deforestation and soil erosion. But the damage caused by it is apparent only because most of the forest has been devastated for commercial purposes. Further, on land that the Adivasis cleared, important fruit and other trees were not cut down and the land was not ploughed up: on the other hand, seeds were dibbled in. "Modern" cultivation may be producing far more erosion with clear felling and deep ploughing. It is also likely that shifting cultivation has been given a bad name in order to force those who practised it into a taxable system.

The mass of revenue

The British, unable to extract revenue from a barter economy, coerced the the Warlis into the cash economy by developing, in the middle of the nineteenth century, a land policy that enforced settlement. Plots were carved out of the Warlis' jungles and legal titles to them given freely. However, the Warlis, not concerned with merely possessing property, took only as much as they needed for their sustenance. The remaining plots were grabbed by non-Warli speculators, who prevented the Warlis from grazing their cattle or getting other plant produce from their newly-acquired lands.[2] The privatization of common property gave the rich legal title to amass land which they did not need for their sustenance.

The British then levied a regressive tax which was based on the number of ploughs that the farmer possessed. A peasant with two ploughs had to pay more than a peasant with one, even if the former possessed or cultivated less land than the latter. Moreover, the Warlis had to pay the tax in cash. Jambu says that many Adivasis left the area rather than pay the British levies.

With the settlement, Warlis were denied free access to what was left of their own jungles. They were made to pay fees for collecting produce from and for grazing their cattle on forest lands. If they did not, their cattle were impounded, and could be released only on the payment of further cash.

With the expropriation of the Warlis' jungles, the British began their destruction. They chopped down trees for their navy railways and military requirements. But the British also cleared the forests because they believed that they were the cause of disease, particularly malaria.

As the Warlis' jungles vanished or were enclosed, all the necessities that came from them had to be replaced by items bought for cash – or the Warlis had to do without them.

A further deliberate step in their impoverishment concerned *tadi* and *daru*. *Tadi* is the lightly-fermented sap of palm trees while *daru* is distilled alcohol. In the Warli area, *shindi* (botanical names are given in the glossary) palms grew in large numbers on the borders of fields and on scrub lands where anyone was free to tap them. The sap was drunk mainly for its high nutritive value; it provided calories as well as vitamins and minerals. It was also considered an aid to digestion and of medicinal value. *Tadi* was essential for survival in the hot dry season, when cultivated food and clean water were scarce.

Daru was brewed from the flowers of the *mahua* tree. These flowers were dried in the sun and then stored. They provided a large part of the Warlis' nutrition. But the flowers could not be preserved through the monsoon humidity, which was why they were used to brew *daru*. The *mahua daru* was not as nutritious as *tadi* though it was also considered medicinal.

The British thirsted for revenue from the sale of alcohol. It proved difficult to tax small home-distillation units, which could be operated when no excise officials were around, or which could be hidden in the jungles. So they set up central distilleries in each province and licensed retail shops in each village, enabling them to tax alcohol at a limited number of points. All other production and sale of distilled alcohol was banned.

At the same time they imposed a tax on trees that were tapped. The tax was rapaciously high and was increased by 933 per cent in 30 years (in 1911, the income from tapping coconut palms, another source of *tadi*, was about Rs 15 per month per tree, of which the tax was Rs 9, leaving a balance of Rs 6 to be shared between the tapper and the owner of the tree).[3] Cash was required, whether Warlis tapped their *tadi* trees or bought their drink from licenced liquor shops.

The Warlis then had to find sources of cash. Many of the Warlis grew "coarse" grains like *nagli* and *kodra*, best suited to hilly land, but these grains had no market value. In this way they were forced to raise marketable crops such as rice even where the land was unsuitable. They had to sell the few fowls they reared and the grass which they required for fodder for their cattle.

The moneylenders

Lack of sufficient cash to pay taxes drove them into the waiting arms of moneylenders. Revenue officials helped by authorizing moneylenders to collect the tax from the Warlis; the moneylenders would pay the tax on the Warlis' behalf and then claim exorbitant interest.

To enforce their claims, the moneylenders would take over their paddy – at very low rates – soon after the harvest, leaving people with too little to live on. When stocks of food were exhausted, the Warlis managed to survive on wild foods that they had to "steal" from their own jungles. But during the monsoon, they had little time for gathering because of their need to grow crops again. So, at the beginning of the rains, they borrowed food from the moneylenders which was to be repaid at the next harvest, at the rate of one and a half sacks for every sack of grain borrowed (this practice is still widespread). This worked out to an annual rate of interest of about 150 per cent, and ensured their continued fall into deeper debt. The liquor retailers also became moneylenders, giving people *daru* on credit, which pushed them further into debt (it should be noted that – prior to the introduction of British laws – when the paid-up interest exceeded twice the principal loan amount, the debt was wiped out by law).[4]

Moreover, land, which was till then held communally, became a titled commodity which could be bought and sold and, above all, mortgaged for loans. Officials eased the process of plunder by simply entering the names of the moneylenders in place of the indebted Warlis' names in the land records. There were numerous such transfers of land, particularly in years of drought. Some liquour retailers accumulated as much as 1,000 hectares of land, whereas the average holding of the Warlis was less than two hectares.

Further impoverishment resulted from the visits of revenue, police and forest department officials. During the stay of the British

Collector in a village, at least three Warlis were conscripted to work for his comfort. Every day, one fowl and four eggs were demanded without payment from each household in turn, and milk, fuel and grain had to be supplied free to the retinue of officials. Warlis were forced to labour in the jungles, planting saplings, clearing firebreaks and carting timber. They were not paid for their labour, and if they refused to work, they were beaten up or prosecuted for allegedly breaking forest laws.

The Warlis were thus driven to sell their labour to the moneylenders and big farmers for wages that barely left them alive. Some were forced to work as tenants on their own lands, with most of the production taken as rent. Under such conditions they showed little interest in raising agricultural yields since any surplus they produced was immediately expropriated by the government and the moneylenders.

But the British desired high, taxable, productivity. The low output of the Warlis provided them with a justification for the transfer of their lands to capitalist agriculturists. Wingate, the originator of the land policy, stated:

> The most effective means at our command for preventing the land from becoming the inheritance of a pauper, or at least a poverty-stricken peasantry, is to afford the greatest possible facilities for its conveyance from one party to another; so that when a cultivator becomes impoverished and by his inability to cultivate properly deprives the community of the wealth it is capable of producing, the land may get into the hands of someone better able to turn it to advantage.[5]

By blaming the victims for the consequences of oppression, the enriched British community hoped they could absolve themselves from the role they played in the creating the "paupers".

The British claimed, as our Government does today, that the taxes were used to provide benefits to the tax-payers themselves. But figures give the lie to that. In a typical village, the total tax extracted was Rs 2, 785 a year, while the benefits received in the form of officials' salaries, education and health services was only Rs 740.

THE SUSTAINABLE TRADITION

Warlis have survived for millennia in harmony with their environment and without oppressing others. Their culture incorporates the spiritual and the material, the living and the non-living, into one integral whole. They consider themselves a part of living nature and hence nature is not exploitable. These holistic concepts may not be verbally expressed as such but are evident from their myths and daily life.

Their story of creation starts with the original self-created couple, Mahadev and Ganga-Gauri, who sowed seeds on earth. Life then began to proliferate so fast that the branches of trees became entangled and the horns of cattle locked. The resulting burden on Dhartari, the Earth, was too much to bear. She sought help from the God Pandu who, however, repulsed and insulted her. This made Mahadev – the older brother of Dhartari – very angry, and he punished Pandu with the kingdom of Death when it came to the distribution of domains. Pandu tried to escape Death, but he was pursued relentlessly, and finally died. Thus Death came to human beings because Mother Earth had been insulted.[6]

After Dhartari, the most revered goddesses of the Warlis are Kansari (Corn Goddess) and Gavtari (Cow Goddess). Land, and cattle to till it, are essential to produce the grain for their survival, and they are both treated as sacred. The jungle, on which they depend for many of their basic needs, is revered in the form of Palghat, the Goddess of trees and fertility.

Their concepts are a reflection of a reality that has enabled them to survive for a very much longer period than the few centuries arrogantly conceded by rational Western science. The perception of the Warlis instills humility, that of the West licenses the domination of nature and of other peoples, and takes pride in such domination.

The conservation of many plants and animals in their jungles is a part of the Warlis' culture, embedded in and perpetuated by customs and religious beliefs. The sacred groves of the Adivasis and Hindus have turned out to be the few remaining areas in India with climax forests and wide species diversity, since no animal or plant could be harmed in them. Their respect for all life reaches down to the smallest creature and plant.

The Warli mango song, chanted at weddings, vividly describes the place of each part of this tree in the ceremony, with a chorus for emphasis. Other songs deal with the uses of and relationships with plants and animals. Their musical instruments are made from plant materials, the principal being the *tarpa*, formed from a dried gourd. Warli dances have the whole village holding hands and moving together in unison.

An important tradition is the painting of Palghat on the mud walls of their huts. Small paintings on cloth or paper, with a background the colour of mud, depict the animals and plants of the jungle together with scenes of their own daily domestic, agricultural and recreational activities. Larger ones represent myths, fables or stories created by the artist.

Co-operation and people's participation are often promoted as if these were recent discoveries of social scientists. For the Warlis, acting together has enabled the community to survive for centuries. The differentiation of labour – and its commercialization – is claimed to be a sign of progress. But Warlis share their labour without charge whenever required for agriculture, house building or any other activity, and by doing so have avoided the economic class-stratification which differentiated labour promotes.

An Adivasi from North India was asked what they do with unwanted children. She replied: "We have no unwanted children. If children lose their parents, they are lovingly looked after by the community." Warlis, too, do not have unwanted children: whether they have lost their parents or are born of unwed mothers, all are accepted with love and affection. One Warli, Savitri, said: "If there is less food, the adults in my family may go hungry but our child is fed properly." Warlis punish their children very reluctantly, as they look upon this practice as barbaric, which could be one reason why many reject the formal education system. Krishna Vavre gave up school after being beaten for not studying.

There are no Warli beggars: the handicapped, as well as others who are incapable of fending for themselves, are all cherished. Though some Warlis may own much more land than others, they all live in the same style in simple huts, without ostentatious show of wealth. Those who earn more tend to share it with others rather than save it in a bank or spend it on accumulating material possessions.

Warlis have a different concept of time from that of urban, westernized people. The West sees time as linear and continuous. This readily generates ideas of unlimited improvement, indeed, progress, which are then harnessed to the need for more goods and services.

As with other peoples of the East, for the Warlis time is cyclic. Though Warlis are often on the borderline of subsistence, this concept of time enables them to survive without undue stress or anxiety about the future.

Their mental security frees them to work for a little while and stop and enjoy life when they have enough for the moment. The pace at which life moves is gentle and relaxed. Those in the mainstream, however, misinterpret it as laziness, lethargy and resignation. The absence of the desire for private gain and the practice of co-operative labour are seen as barbaric, while unlimited greed and the oppression of workers are civilized.

Warlis are not saints living in forests. They have the same faults and frailties of all humanity, but their system is free of many of the flaws which have been intensified by the spread of Western monoculture.

Warlis are careful not to interfere unduly with the environment. As Kings of the Jungle, the Warlis act as its caretakers rather than exploiters. Nature responds, providing them with a wide variety of goods for their survival and much else besides. The generosity of nature is not abused by excessive consumption. Their contentment with a basic sufficiency is interpreted by the mainstream as an inability to comprehend the value of formal education and an unwillingness to absorb Western technology, both considered essential for their progress.

Global climatic changes are now disturbing the rhythms of nature. The Warlis may not connect the protests of nature with the mayhem of humans, since the sites of destructive activity are remote and the links invisible. In the degraded landscapes, the Warlis may come to believe that their ancient traditions have failed them: their faith in nature may be shattered by the activities of economically rich and powerful – though distant – strangers. The contamination of alternative sustainable practice could be the final legacy of a world-system that has belatedly declared its own dedication to sustainability, at the very moment when it is willing the extinction of actual living examples.

Knowledge

Many Warlis do not know how to read or write but they have a vast store of knowledge, orally handed down from parents to children for countless generations. Yet the dominant society equates their illiteracy with ignorance. They have no need to write things down since, without instant access to their knowledge, they would not survive in their demanding environment. Their memories have expanded extraordinarily to store it reliably. Each individual has to be an expert in all aspects of knowledge, with very limited amount of specialization. This, together with little division of labour, and a conscious effort to be self-reliant and self-sufficient, contributes much to equity.

Raji Vavre, Krishna Vavre's elder sister, a twelve-year-old girl, knows the names of over a hundred herbs, shrubs and trees and their varied uses. Many of these supplement her basic diet of cereals and pulses with essential proteins, vitamins and minerals. She knows which plants are a source of fibre, which are good for fuel and lighting, which have medicinal uses. She knows how to get crabs out of their holes and how to trap fish. She can catch wild hare, quail and partridges and locate birds' nests.

She possesses a vast, complete knowledge system, which includes such orthodox divisions as animal husbandry, agriculture, meteorology, herbal medicine, botany, zoology, house construction, ecology, geology, economics, religion and psychology. And, more important, she is also part of a remarkably successful educational system: because her father has died, she taught all this to Krishna and her other younger brothers and sisters. In this system there are no drop-outs or failures, and the educated live contentedly without relying on a multinational company to give them a job. Moreover, Raji is not unique. Most Warli children of her age have the same or even more knowledge.

Of course, not all the knowledge of the Warlis is valid. There is no doubt some superstition involved – and a few of their practices are positively harmful – but this is a tiny fraction of their "science". And they have the wisdom to use their knowledge well.

The erosion of knowledge

Unfortunately, an increasing number of Warlis, particularly those who have attended schools, have not retained their heritage. The formal educational system devalues their knowledge. Interactions between the formally-educated and the Warlis can be revealing.

A postgraduate in one of the life sciences asked a passing Warli the name of a rare, beautiful, flowering tree. The Warli said it was a *bhendi* tree, to which the graduate replied, "impossible, I know the *bhendi* tree well", referring to the common *bhendi* tree, *Thespesia populnea*. Some time later we collected the fruit and leaves of this same tree and showed them to Janiya Ghatal, who has had no formal education at all. He looked at them carefully, felt the texture of the leaves, checked their smell and taste, and, after a little while – as if his senses had jogged his memory – said, "This is from a *chera* tree". He also mentioned that it was rare in this area and that there was only one growing about 15 kilometres away – the same one which we had seen. We looked up the name and found that he had identified it correctly as *Erinocarpus nimonii*, one of whose names was also *bhendi*, a name which is used for a number of shrubs and trees all belonging to the *Malvaceae* family. The first Warli must have immediately begun to doubt his knowledge, but Janiya's was reinforced when we showed him an illustration in a book and said that he was right.

The formal health, agricultural and other sub-systems imposed by the dominant groups also aid in eroding Warli competence. These are the major causes of the loss of Warli self-respect, and this facilitates their domination by rich farmers and others.

Further contribution to the decline of their knowledge is made by the loss of common resources in the jungles and "waste" lands, by environmental degradation or enclosure. The variety of species available for observation and use is reduced. Plants need to be seen frequently in order to be recognized; many qualified botanists are unable to identify in the field plants they know only through textbooks and herbaria. Plants, as well as animals and insects, have to be observed throughout their life cycles. But such practical learning becomes more difficult as the Warlis' habitat is taken over by factories, power stations, dams and roads.

Such knowledge may be essential for all of us if we are to build a just, sustainable society. The illiterate – Adivasis as well as non-Adivasis – who have preserved traditional practice may now be revealed to be superior to the literate, who have been converted by education to the superior reason of injustice. It could be that the illiteracy rate in a region is an index of the sustainability of that region's systems. Writing and reading, instead of serving human liberation, have contributed more to oppression and the spread of an unsustainable monoculture. This is a result of the control imposed by those in power over sources of information. Literacy in such a context itself becomes an instrument of subjugation.

One way of preserving their wisdom is to learn from them and return to them what has been learned. This reassures them about the value of what they know. While classroom learning can never take the place of their oral system, it is becoming necessary now to record their knowledge, not for the purpose of commercial exploitation, but for strengthening that knowledge, since much could be lost forever.

Even if their knowledge were to be included in the curriculum of the formal system, the Warlis feel that their children learn more by tending siblings and cattle, helping out in cultivation or merely accompanying them in their jungles. All these constitute true education, with children learning a sense of responsibility as well as an ability to observe relations in nature. Watching Krishna tend his herd is instructive. As he drives them to the *nalla* for a drink, he darts to look at a flower, observes a bird building its nest, tries to catch a dragonfly, picks *karandi* berries, exchanges information with his friends similarly occupied, bursts into a song of his own composition and interrupts it to shout at a straying cow in a language that it seems to understand.

It is rather strange that, with all the Warlis' research and foresight, those who promote formal education still insist that it is necessary in order to teach the Warlis the advantages of delayed rewards. Warlis, however, look even further than the achievement of a certificate which does not even guarantee employment to its holder. They balance the small probability of getting a job after such education against the loss of traditional self-sufficing knowledge, and schooling loses out, since in their eyes the educated unemployed have been made truly ignorant.

Moreover, the normal school is an urban product, with an urban timetable that is unsuited to agricultural needs. Literacy and the school need to be adapted to the life of the community, not the other way around.

Crops

Their agricultural system is almost always called primitive. It is in fact sophisticated. It has been developing over several millennia by a close study of the interactions between plants, animals and human beings. If it has survived for so long, it can only be because it is ecologically sound. The practice of traditional agriculture does not mean rigidity in techniques; rather, it is a way of thinking which maintains a dynamic equilibrium and which can be adapted as changing circumstances require.

Their priority is to grow enough paddy and pulses for their own consumption, but they grow oilseed and fibre plants as well as vegetables in an attempt to be self-sufficient. They have evolved a complex method of multiple cropping which appears unique. Dependent as it is on a short, irregular, rainy season, it maximizes food security, makes optimum use of the available resources of land, labour and animal power.

Warlis usually cultivate a few of more than 15 major traditional varieties of paddy that are available to them. These have differing requirements of water and soil, mature at staggered times, have different susceptibility to pests, and possess varied flavours. If the monsoon proceeds normally, their main varieties give a good yield; if it doesn't, the others, more hardy but with lower yields, still produce some food. While their varieties may not give a higher maximum yield than a new High Input Variety (HIVs, also called High Yielding Varieties) would give in a year with good rain, HIVs yield little or nothing in an even slightly abnormal year. The average yields over a number of years could, therefore, be higher with their traditional varieties.

Some varieties grow on land that is bad, hilly or that has lime in it. One black variety is harvested within 45 days of sowing. Most paddy varieties require to be stored for some time before use but this one can be consumed immediately, thus giving them food soon after the start of the rains. Such varieties may be essential for all of us if the monsoon becomes more erratic because of global warming.

Warlis have experimented with HIVs, which they claim impoverish the soil. So, if they plant HIVs at all, they rotate them with traditional varieties which, they say, restores soil fertility. This shows that Warlis are still carrying on field research with long-term planning in mind.

They follow the same system of transplanting as other farmers in the Konkan region (see Chapter 5). Manuring is carried out using leaves and twigs from their jungles, a small area of which can provide sustainable yields, since the trees are never destroyed and their jungle does not have to start from scratch every cycle. Further, they claim that they lop branches only from trees whose boles will be improved by lopping. In addition, paddy straw that was used in the previous year's thatch and twigs from old fences are also recycled. All these are burnt, together with some cow dung on the seedling fields. The manure used for the main-crop fields is also organic.

Warlis say that if they use synthetic fertilizers they will get a good crop in the first year of use, but that in subsequent years the yield will go down, and they will need increasing quantities every year. We know now that this happens because the inorganic nitrogen lowers the output of nitrogen-fixing plants. The replacement of organic matter by synthetic fertilizers also reduces the ability of the soil to hold water and, as they say, "the soil gets dry". This makes the crop more susceptible to erratic monsoons as well as to erosion by wind and water. So, even though synthetic fertilizers are provided to them by the government free or subsidized, they do not use them, or use them only for a year or two and then switch back to organic manures. Curiously, Adivasis in other areas cultivate in the same manner. On the other hand, numerous "educated" sugar-cane farmers in Nasik district have continued to use such fertilizers until nothing at all can grow on their soil. They then find it difficult to revert to traditional practices.

Some Warlis plant *dhedhar*, a species of *Sesbania*, as a green manure crop in the main fields just after the rains fall. This is ploughed in after about three weeks, when the field is puddled immediately before transplanting. The seeds are obtained from wild plants or those growing in hedges. A similar species, *Sesbania rostrata*, has very recently been tested out by the International Rice Research Institute in the Philippines, and is being used in a similar manner, but with irrigation. The Warlis had discovered this technique years earlier.

Warlis integrate paddy with a number of other crops and manage to get two harvests from the same plot without irrigation. They sow *tur, jowar* or *chaoli* and paddy in the same seedling beds. On the borders of these beds, they plant *ambadi, lal ambadi* or *bhindi*. After the paddy seedlings are removed for transplanting, the rest thrive on the residue of nutrients not taken up by paddy.

After the quick-maturing paddy is harvested, *tur* and other crops are planted on the same fields. Other monsoon crops are an oilseed, *khorasni*, and a pulse, *udid*. The best paddy varieties grow only in good soil which retains its moisture. After their harvest, *harbara, val, rai* and other crops are grown in these fields since they survive on the residue of moisture in the soil. The legumes help in maintaining soil fertility. Where low-lying land stays damp after the monsoon, a second crop of paddy is often sown.

Some Adivasis intercrop several plants. In the manure in between rows of maize, cucumber and *bhindi* are cultivated. *Ambadi, khorasni, udid* are cultivated on the border of these rows.

Several other field crops are grown, while some vegetables such as *palak* and *loni* and trees such as *shevga*, mango, *bor* and other fruit trees are grown in backyards. Cucumbers and pumpkins are trailed over their huts.

Warlis rarely use chemical pesticides, because of the harm they do to the paddy ecosystem. Besides, they have their own methods of pest control. When paddy was attacked by the brown spot disease, in order to control it Lahanu Rade suggested putting the leaves of a wild tree, *khair*, in the channel through which the water flowed into the field.

Warlis "plant" branches of trees in their fields for birds to perch on when they search for insects, but since they use those of certain tree species – *shindi* and *kuda* – only, it is possible that these also possess some pesticidal action. In fact, probably because of their organic methods of farming, there is little damage by pests. However, the excessive use of synthetic pesticides by neighbouring non-Adivasi farmers is destroying predators as well as creating pests out of normally harmless insects, which sometimes attack their crops with disastrous results.

Pest-resistant crop varieties are also chosen. The variety of *harbara* grown has small pods unlike that usually cultivated as a cash crop by non-Adivasi farmers. It has been found that this latter variety is more susceptible to pests.

Paddy requires good soil for its growth and many Warlis own land on hillsides which is rocky and infertile. But even on these they grow crops like *nagli* and *kodra*. Some species of kodra are toxic but they know of ways of detoxifying them. Modern agriculture has still to find better methods of using such marginal lands.

However, rice is considered a superior food by the encroaching "civilization", even though its nutritional value and yield are less than those of the so-called coarse grains. Some Warlis have stopped growing the latter with a consequent loss in food security as well as nutrition.

Their agricultural practices have the aim of ensuring sustainability even at the cost of low output and immediate income. This is exactly the opposite of what the Western system promotes: maximize income now and don't bother about the future. But the Warlis' attitude is normal for people who know that they have to live within a specified region for centuries: sustainability has to take priority. It is the "civilized" who fantasize that they are free to escape their local ravaging by migrating to cities or other countries to exploit still undamaged environments.

The monsoon season, June to October, is a time of hard work – sowing, transplanting, weeding and harvesting. They do not have time for celebrating any feasts in this period. Threshing is onerous but the introduction of threshing machines, which will be appropriated by rich farmers, will also reduce employment. Husking paddy is also strenuous for women, but in the mills the rice is polished, thereby losing much of its nutritive value, whereas hand-milled rice is never polished. In late October and November come the feasts of Dassera and Divali, which are celebrated with joy and dancing. In January they begin preparing for the wedding season which is in February. And after that, they start collecting manure again.

Water

A person from Bombay bought a piece of agricultural land in a Warli area and called in an expert water-diviner to point out locations for wells, at a cost of Rs 500 for each. He then drilled a borewell at one of the spots, going down to more than 20 metres, for which he spent about Rs 22,000 and got not a drop of water. Lahanu, his employee, then suggested that he dig open wells at two places which he indicated without charging him a *paisa*. At both the spots he got ample water at a depth of about three metres, one of these being the only source of water in an area of 12 metres diameter. In another case a pit dug within six metres of the point indicated by Lahanu was absolutely dry although it was nearer a *nalla* than the one in which plenty of water was found.

Lahanu and other Warlis locate water by observing land crabs. The mounds of wet mud that crabs excavate and leave around their holes indicate the presence of water below. The larger the number of crab holes at a given spot, the greater will be the availability of water. In one location, the well diggers reached hard rock at about four metres and gave up hopes of finding any water there. But there was one crab hole going down through the rock and Lahanu told the diggers to proceed. Beneath a 50 centimetre layer of rock they found soft earth and plenty of water. It's as simple as that; no incantations or mantras, no divining rods, forked twigs, not even magnetic needles or resistivity measurements.

Warlis use only a little water for irrigation, perhaps aware that if all irrigate their fields, the water table will be depleted so rapidly that in a few years none will have enough. This has happened in the neighbouring "civilized" areas of Gujerat.

Animal husbandry

The Warlis' attitude towards animals – and life in general – is illustrated by a story told by Chandrakant.

> You know, there was this Warli who wanted to sell his bullock. He approached a farmer, both chatted for a while, and then negotiated the price. When the deal was fixed, the seller offered a *bidi* (a cigarette made from the leaves of the *temburni* tree) to the buyer, who however, politely declined it. The first Warli said to the second, "What? Don't you smoke? Then I won't sell my bullock to you"! He then went to another farmer. Again, after the price was agreed upon, the first Warli offered a *bidi* to his customer. But this time, the buyer accepted it and our Warli

was happy to find someone who enjoyed smoking. Do you know why? Why did he not sell the bullock to the first customer and why was he so happy to find a customer who enjoyed smoking? Because a person who does not smoke is a person not willing to relax and if he, himself, is not willing to relax then he may not allow the bull to relax either.

Cattle are essential for ploughing and traction, hence their importance. The residues of most crops can be fed to cattle. Two bulls are normally required for ploughing, but those who have only one bull share it with others in a similar situation. Tractors do not produce any fertilizer, are expensive, damage the soil and would require the bunds of their small fields to be broken down and rebuilt for every ploughing.

Warlis do not drink milk since their normal diet is nutritionally complete; they maintain cows solely because they produce bulls. All the cow's milk is usually left for the calf, since Warlis feel that the calf has a right to it. Their cattle are housed in the same huts in which the people live.

The bulls also serve to draw carts which are the only possible means of transport across rough field tracks and unevenly-metalled roads. Constructing good roads to all the hamlets would remove from use a lot of fertile agricultural land, and would tempt outsiders to enter the area and buy up more.

Another example of their sensitivity is their attitude towards cattle that can no longer work. Ladku Vavre, Raji's grandfather, says that animals which have worked hard for years should not be sold for slaughter since they are entitled to a leisured old age. They are, therefore, free to move about as they please. This is considered wasteful by the system since they consume scarce fodder and they would be "better used" for meat. But such commercial butchery would reduce the manure available and would also deprive other Adivasis, who skin dead animals and collect their bones, of essential income. Further, if the animal is sent to the city for slaughter, the bones are bought by multinationals like Hindusthan Lever and exported, which is a loss of precious phosphorus fertilizer.

Warlis are able to diagnose several cattle diseases and have herbal remedies for them. Sores on cattle are treated with the leaves of *sitaphal* or *karela*, the seeds of *palas*, or the tubers of *kovli bhaji*. Cattle suffering from worms are treated with *hing* dissolved in water. Foot-and-mouth disease is not very common in local breeds, but when it does occur, Warlis say that it is only necessary to feed the sick animals with molasses, since the animals die of hunger because their mouths get sore and they are unable to eat.

Warlis rear poultry, guinea fowls and pigeons, feeding them mainly on wastes. These are also treated with herbal medicines. For fowls suffering from diarrhoea, for instance, the bark of *kuda* is powdered and put in the drinking water.

Free food

Newspapers often report that in times of scarcity, Adivasis "grub in the forest for roots" to live on. Jivya Mashe tells the story of Warlis who asked for food from a rich farmer in a year when their crops had failed. The farmer refused, saying, "I have just enough. I am sorry, I cannot help you". To his surprise, the Warlis replied quite cheerfully, "we understand. You don't have to worry. We have bearded farmers for friends. They will provide us with food." And they went into the jungle. The farmer followed them, curious to see who these friends were. And he watched as they moved through the jungle pulling up plants from the ground. When they held them up he realised what they had meant. The plants had long, hairy roots.[7]

These roots are the tubers and corms of yams, *suran* and other species. These monsoon plants are either climbers or herbs. Each tree and shrub in a forest will have one or more of these climbers, some rising only a metre or two, while others reach the tops of the highest trees, filling all the microniches available to them. The tubers grow larger if undisturbed for a few years. In the dry season, if there is a shortage of normal cereal grains, Warlis dig them out. They cook them in a special way since they usually contain poisonous substances. It is certainly trouble to detoxify them, but this is the price they have to pay for free food production and storage. The toxic substances prevent rodents and insects from attacking the tubers.

In addition to tubers, Warlis get a wide variety of food from wild plants. The mainstream is convinced that these free foods are inferior and that Warlis eat them because they have nothing as good as commercialized foods. But this is far from the truth. Most wild foods are nutritious, tree-ripened, and delicious. Hunting provides a further range of foods. All these supplement, right through the year, the rice and pulses that constitute their staple diet.

Warlis do not need to search for these as the locations of the various plants are all stored in their memories, together with information on when they can be harvested.

It is interesting to speculate whether the first humans who cultivated crops were seduced into doing so by claims of a Green Revolution using modern, high technology, just as farmers today are being conditioned to move in the same direction of higher labour and other inputs. All such processes serve only to provide an extractable surplus for rulers, urban dwellers and industrialists.

One would think that common resources would be seized by the most powerful in a village, and while this happens in some cases, Adivasi traditions of co-operation discourage it. Among the tuberous plants which spring up with the first monsoon showers is *kovli bhaji*, whose leaves are eaten. But Warlis pluck them only on one particular day, thus ensuring that all who want them can have them, while at the same time over-exploitation of the species is prevented.

A month before the monsoon starts, they collect and eat the flowers of what looks like a larger version of *kovli* which they call *kovli-cha-mama* (kovli's uncle). The bulb itself, used in allopathic medicine, is also used medically by Warlis. At about the same time a climber, *koland*, emerges whose stem provides drinkable water in the hottest season.

From then on there are a large number of various edible mushrooms, herbs and climbers whose leaves are edible, all available for the picking. During the latter part of, or after, the monsoon, plants produce edible flowers, fruit and tubers. During the dry season the larger shrubs and trees fruit. The dried flowers of single wild tree species, *mahua*, formerly supplied over 30 per cent of the Warlis food requirements. The berries of *karandi* form an important part of their diet, pickled raw or eaten ripe. Well-intentioned voluntary organizations promote the pickling and preserving of these for sale in cities, but the cash obtained may not compensate for the nourishment lost.

Janiya knows about a hundred plants which are edible and consumed by most Warlis. Some, normally considered harmful, are eaten in strictly limited quantities or cooked in ways that render them edible. It is possible that many of these wild plants may, if cultivated, yield more, provide better nutrition and better food security than the presently cultivated species.

Ragu was walking along a road through a jungle when a *ghorpad* (monitor lizard), more than a metre long, ran across it. The flesh of these reptiles is a favourite food but catching one by chasing it is next to impossible. Ragu simply gave a peculiar low whistle and the *ghorpad* stopped moving and waited to be caught. Sometimes, the leaves of *dudhkodi* are burnt at the opening of a *ghorpad's* lair in the ground to smoke it out. Ramji Kakad has trained his dog to sniff them out.

Delicacies like quail, partridge and peacock, besides common birds like waterhens and lapwings, are caught by throwing stones at them. Yeshwant makes birdlimes out of the sap of wild fig species. He makes several kinds, some with a short effective life and others with a long one. He coats twigs with the sap mixture and heats them until the consistency is right. He then makes a trap with the twigs, places it on a convenient tree and baits it. The next time he passes that way, he has just to pick up the bird caught in it. Eggs are also picked off when passing. Bunting and lapwing nests are found while ploughing the fields. The Adivasis hunt solely for dietary purposes, never for the pleasure of killing.

In many of these hunting activities, the co-operation of neighbours is essential – and also more enjoyable. Sawant Vilat heard a chicken squawking in the drowsy time after lunch. He rushed out to see a mongoose running away with it. Without letting the mongoose out of his sight, he called out to his family and neighbours. They chased it from one side of a *nalla* to the other in a well co-ordinated manner derived from long practice, until it was exhausted and caught. The only weapons used were sticks and catapults. It was killed to be eaten and is about the biggest wild animal that is still available in this area: deer and other game have been eliminated by non-Adivasi sportsmen long ago. Mongooses kill their fowls, forced to do so because their normal prey of snakes and rats are also being reduced by habitat destruction.

In hunting hare and partridge, one Warli will climb a tall tree in sparse forest land. The others will beat the bushes and when the prey is seen moving, the one on the tree directs the others to surround and catch it.

Some of these birds and animals are protected species, but the Warlis have a right to them since they need them for food, not for fur or feathers. Further, since they keep their needs to a minimum – no eggs for breakfast every morning, no partridge for dinner every Sunday – the Warlis are not responsible for any diminution in their numbers.

About ten species of fish live naturally in flooded paddy fields and *nallas*. Warlis have developed several methods of catching them. The technology they use, as usual, requires a minimum of labour. And again, much of it is done as a leisure activity. No expensive inputs such as nylon nets are required and hence they do not need to sell any of their catch for cash.

Years ago Ladku Vavre had built two cylindrical stone pillars about a metre in diameter and two metres high, on either bank of a small *nalla*. He did not use mortar but bound the pillars with ropes made from the bark of trees, which do not rot in water. The pillars have withstood the force of strong monsoon currents without any failure. He has a screen of bamboo slats with spaces between them, framed in strong wood. At the beginning of the monsoon, Ladku ties the screen firmly across the pillars on the upstream side. It extends from the bottom of the *nalla* to about a metre above the highest water level, at a slope of about 45 degrees. A screen lasts for several years. He works on these before the monsoon, when he is relatively free. During the monsoon, heavy rains wash fish out of paddy fields into the *nallas*. The monsoon is a busy and tiring season, but Ladku and his grandchildren have only to sit on the top of the device and pick off the fish which are forced up above the water. Small fish pass between the slats.

In a small dam across another stream, Janiya made a channel through which the water flowed. In this he fixed a conical net made of *ambadi* fibres. He then put the powdered bark of the *inodi* tree near the mouth of the net. Fish are intoxicated by the bark and are unable to make their way out of the trap. Janiya has caught fish more than 30 centimetres long in his trap, getting them daily for several months in the year. When the *nalla* level goes down, the fish try to go downstream by jumping over the dam; Janiya invented a simple chicken wire device to trap them.

In May, people from the neighbouring Patilpada come to catch fish in the *nalla*, just before it dries up. A group of four or five of them drags a sari along the bottom of the *nalla* with their feet, holding the other edge above the water level. A number of groups co-operate, all moving together towards the bank or the area where fish have been sighted. The fish are not shared equally; whoever catches, takes, but all is very friendly and entertaining.

Fish are stupefied by throwing leaves, fruit, roots or bark of any of several plants into the stream. The most commonly used ones are *ghela*, *beheda*, *hed*, *karanj* and *alu*. The plant chemicals do not have any toxic effect on humans.

Ramji catches freshwater prawns up to 20 centimetres long with a remarkable efficiency of labour. He goes along one bank of a *nalla* feeling out, with his toes, the holes that prawns make below the water level. He puts a mixture of the bark and leaves of three plants in the holes and after covering about ten metres, returns to do the other bank. Then he goes around once again picking off the stupefied prawns which come out to the entrance.

Gopal Rade dives for eels which normally lie under stones, when having his evening bath. Turtles are caught as they cross land.

But "progress" is reducing food availability, directly by the destruction of the forests and also indirectly. Prawns breed only in brackish waters at the mouths of rivers. The fry then swim back up rivers and *nallas* during the monsoon. Big dams prevent them from going up river, with millions of fry dying downstream of the dams. And with the destruction of the watersheds, small streams dry up soon after the monsoon.

Instead of using plants for stupefying fish, some are tempted to use dynamite, although it is illegal. This destroys all the creatures in the *nalla*, and often harms the fisher, too. Balu had his right arm blown off because he did not throw the dynamite fast enough. Then there are the modern city-based farmers like Narayan whose father is the managing director of a large tobacco company. He pours a highly toxic pesticide into the *nalla* to kill fish, regardless of the possibility that the fish he will eat can give him cancer and other diseases. Worse, these pesticides affect thousands of others downstream.

The loss of the commons and of free food is a loss of freedom of choice on a more fundamental level than any gain in freedom of choice (between, for instance, different brands of tomato soup or cakes of soap) which is offered in its place.

Housing

The homes of the Warlis are small, simple structures of mud, reeds and thatch, renewable materials only. They have few windows and are dark and cool inside, appropriate for this climate. Any operation that requires light can easily be done out of doors. They see no need to build ever-larger homes even though materials are freely available. They use little or no furniture.

Several types of timber are employed, according to the differing stresses and strains to which they are put. The best are *saag, inodi,* and *shisham,* with *hed* and *asan* following. The preferred reeds come from a shrub called *karvi,* which grows on the surrounding hills and flowers once every seven years. The reeds are bound together with cords made from plant fibres and plastered with mud mixed with cow dung. This makes the mud waterproof and is not washed off even by heavy rain. The floors, also of mud and cow dung, absorb any water that spills on them without getting soggy. They are normally swept very clean. The interiors are fragrant and inviting. The thatch used is paddy straw. Thatch and reeds need to be replaced regularly but all the used materials serve as manure.

Warlis find it increasingly difficult to get the materials they need. The best timbers have all been cut down by commercial interests and Jaya complains that they have now to be content with what they would not normally touch: *shimti, kakad* and other weak woods, or those which are susceptible to white ant attack. The widely promoted *kubabul* is not used since it quickly rots in the soil.

Karvi is also becoming scarce and they are often compelled to use cultivated *jowar* stalks which do not last long and are attacked by insects. Paddy straw, termed an agricultural waste, is now being used for paper and boards, making it unavailable to those who do not grow enough of their own.

Many of the Warlis now use roofing tiles instead of thatch as this is considered a status symbol. Further, electric connections are not given for domestic use unless the hut has a tiled roof, because of the supposed danger of fire. Tiles reduce the labour required to replace straw every year. Their disadvantage is that they make the hut hotter, because much of the solar energy they absorb is radiated inwards. This increases discomfort and may lead to more disease because of heat stress as well as the spoilage of food. Those who have tiled their roofs now still have to put a thatch over it to keep it cool. Tiles use up precious top soil, since good clay is not available in the area.

Rich farmers now build houses of brick and some Warlis imitate them. Vikas has sold his fertile top-soil to a depth of half a metre to a brick-maker. This soil has taken hundreds of years to build up, but he is not worried because heavy rains will wash in the soil from neighbouring fields, which means that yields over a wide radius will be diminished. The loss of agricultural soil because of brick manufacture could reduce food production considerably.

The manufacture of baked bricks requires a lot of fuel and with the increasing scarcity of fuelwood, paddy husk is now being used. With husk acquiring a market value, rice mills no longer return it with the rice to the owner. But paddy husk was composted or burnt in the fields and so another reduction in fertility is promoted by modernization. And baked bricks cannot be recycled.

Energy

Firewood trees on jungle lands around the villages have been almost totally cut down. Savitri Vilat now spends up to half a day, several times a week, collecting fuel for her daily needs. Fuelwood is normally taken in the form of side branches of certain trees, which preserves the main trunk.

As forest destruction continues, the quantity easily available is reduced and so is its quality. Women are very particular about the type of firewood they collect. Mathu says that she prefers the wood of *inodi*, *saag* and *karandi*. During the monsoon, she uses twigs of *shisham* and *likandi* that burn well even when fresh, but if these are not available, she has to store enough fuel for nearly four months. She does not use *temburni* because it burns so fiercely that her earthen cooking pots crack. Neither does she use *savaar* or *kahandol* because they attract insects. The latter also smokes too much.

If wood that burns well is not available, women spend more time in attending to the fire. If the wood smokes excessively, they suffer from burning eyes and sore throats. The smoke has also been shown to be highly carcinogenic. The types of trees planted under social forestry programmes to solve the fuelwood problem are selected for fast growth and high calorific value and do not take these factors into account.

The shortage of fuel could also mean that yams and other foods may not be thoroughly detoxified or that food may not be adequately cooked, contributing to nutrition deficiency.

Biogas plants are economical only if the owner has at least four head of cattle, which few Warlis do. Those who have put up biogas plants employ people to collect dung from roads and "waste" lands, thus reducing the free manure available. On the other hand, the use of biogas by the rich results in fewer trees being chopped for fuel.

When he has to go out on a dark night, Chintu Vilat impales a number of kernels of *diveli* seeds on a thin twig and has an instant torch. Oil was formerly extracted from these and several other seeds for house lighting. But now it is more economical to sell the seeds and buy kerosene, a non-renewable resource priced absurdly low.

Fibres

Warlis use the bark fibres of many plants for making cords and ropes. Some climbers – such as *chai* – are used directly for tying, this one for binding thorny branches into hedges. The most commonly cultivated fibre plant is *ambadi*, but ropes which have to be used for special purposes are made of fibres from particular wild trees. Janiya plants just enough *ambadi* for his own needs. He uses some of the plant's leaves for food. After harvesting the stems, he submerges them in a *nalla* for retting for a few days. He spins the lustrous fibres, using a small spindle made out of hard, heavy *shisham*. He spins in the evenings after work, producing a few metres of cord per minute. He also makes multistrand ropes up to three centimetres in diameter for cattle harnesses and other purposes. After the ropes lose their strength, they are recycled for manure. Nylon, the modern alternative, requires cash for its purchase, is based on non-renewable mineral oil for its manufacture, is not biodegradable, and disemploys those who make ropes from vegetable fibres.

Ambadi fibre has been found to be excellent for paper manufacture. Farmers are familiar with growing *ambadi*, it does not require irrigation and is harvested within six months of sowing. Yet the government is encouraging farmers to plant exotic eucalyptus on their best irrigated land, with the first harvest obtainable after only five years.

Health

The equivalent of a doctor among the Warlis is the *bhagat*. *Bhagats* are, however, more than doctors: they also retain and transmit the myths and legends of the people.

On Bhadrapad Vadya Dwadashi (the twelfth day of the second fortnight of Bhadrapad) at a big *mela* (fair) held exclusively for the Warlis at Mahalaxmi, *bhagats* and those who aspire to become *bhagats* get together. The *bhagats* get into trances and try to communicate the skill of getting "possessed" to the aspirants. Their actual annual training sessions are held during the time of transplanting paddy, with shorter camps from Dassera day on till Vaghbarus, two days before the main Diwali feast.

During the day, they work on their fields as usual, but in the night they get together at the place fixed for instruction. Aspirants avoid all contact with women during this period. Women are not even allowed in the vicinity since the Warlis are suspicious of *bhutalis* (witches). These workshops are called *ravals*, and are generally sponsored by a rich Warli seeking a child.

It is not only the *bhagats* who possess knowledge of herbal medicines; most individuals do also. When women were asked at a regular women's meeting at Talasari what they knew about herbal medicines, they recited dozens of names of plants and their uses. Vasanti Dandekar had far more knowledge than the rest of them, a probable consequence of her father being a *bhagat*.

Among the ailments treated by ordinary Warlis are allergies, anaemia, colds, coughs and other respiratory diseases, toothache, diabetes, diarrhoea, dysentery and other stomach problems, ear and eye problems, fever, fractures, liver diseases (including jaundice), malaria, measles, menstrual problems, migraine and other headaches, piles, stings and bites, skin infections, and intestinal worms. Information on 85 plants used for these has been recorded, but much more remains uncollected. Several herbs are used in combinations that act synergistically.

Over forty plants are used for diarrhoea, dysentery and other stomach complaints. One of them, *kuda*, produces the allopathic drug for dysentery, conessine. Fifteen plants are used for fevers, five for menstrual problems, 13 for colds, coughs and other respiratory diseases. The twigs of five plants are used as toothbrushes. One of these, *neem*, has been found to be effective in the treatment of bleeding gums and pyorrhoea. Others contain compounds that slow the production of lactic acid and dextrans, both of which contribute to plaque and tooth decay.

Women often get fungal infections on their feet during transplanting time. To prevent this, they coat their soles with the pulp of the raw fruit of *temburni*, which is available at just the right time. This pulp was formerly used to waterproof boats and terraces of buildings. Three other plants are also used for fungal infections, including one for thrush.

Other forest produce

Adivasis are now legally permitted to collect bamboos, reeds, seeds, resins, gums and other "minor" produce from their own forests. These are required for their own use as well as for the production of articles for sale. Those with little or no land acquire indispensable income in this way.

Among the oilseeds collected are those of *mahua*, *karanj*, *kusumb* and *diveli*. The oil extracted from the latter is mainly exported to Europe for use as a lubricant in wool spinning. *Inodi*, *palas*, *kahandol* and several other trees are tapped for gum. Most of the seeds and gums are exchanged for items not grown by them such as onions and dried fish, or sold for the cash required for the purchase of clothes and other essentials. There is much scope for processing the raw materials in the villages for their own use as well as for value-added sale. No capital investment is required, the know-how exists with the people and the work can be carried out even in their spare time.

Only the Forest Department is authorized to purchase oilseeds and resins, but the officials themselves often request non-Adivasi village traders to handle this lucrative business for them. The result is that the Warlis are paid much less than the official price with much of the profit going to the middlemen.

Warlis make baskets of bamboo which they need for their farming operations. Bamboo is becoming scarce in their jungles, so many of them grow it in their own backyards, but this displaces the vegetables that they grew earlier. Woven *chatais* (mats) are made using palm leaves. Soni makes a *chatai* of about one and a half metres square using 12 to 15 woven strips, each about ten centimetres in width, stitched together. A *chatai* is sold for Rs15 to 20. Soni claims that only a few Warlis still possess the special skill to make *chatais*. Raincoats, essential for working in heavy rain, are made from leaves and bamboo.

Children's toys are made of forest products, too: whistles are made of the leaves of several plants, the seeds of *sagargota* are used as marbles, and running carts are made of bamboo and reeds.

Warlis rarely go into their jungles for the sole purpose of obtaining one particular item. Several are gathered from February to April when they collect leaves and twigs for fertilizer. They sell them when they go to the weekly market or annual fair. *Palas* leaves are made into cups for holding traditional foods sold by roadside hawkers and must be fresh. A few of these leaves are collected by women when they gather firewood but it would be too wasteful of time and energy for each of them to go to the main village where they are bought by a trader. In Dhindi pada, an old man who obviously could not do any hard work but was able to walk several kilometres a day, collected what each woman had gathered and took them to the trader every day.

THE INVASION OF CIVILIZATION

After independence, the Government did not repeal the British-made forest legislation, with the result that the jungles did not return to the Warlis but remained the property of the Government, to be exploited, again for the sake of revenue. In many areas most of the trees have been completely wiped out while in others only a few are left. While a total ban on tree felling has been recently enacted, the addictive taste for quick money acquired by the timber merchants ensures that trees are still felled illegally.

In the matter of forest produce, too, the immemorial rights of the Warlis are being eroded. Large areas of their jungles are being auctioned to private contractors who in turn employ labourers, often non-Warlis from outside the area, to collect the produce. This not only deprives Warlis of their produce but also results in over-exploitation of their jungles. If too many seeds are collected, too few are left for regeneration of the species and for the use of animals and insects which feed on them. All *kahandol* trees have died out in one non-Warli area, because they were tapped continuously for their gum, instead of being given a rest every few years. There are no trees left to provide seeds in that district. The gum from *kahandol* is exported for use in ice cream and other food products.

Official reforestation programmes also result in a reduction of common resources. It is still standard forestry practice to clear all natural growth of herbs, shrubs and even large trees in order that seedlings can be planted in equidistant, perfectly straight rows. The few species that are planted can never make up for the diversity of what grew naturally, leaving many gaps in the raw materials required by the Warlis. The yams, *kovli* and other tuberous plants are the only ones to survive, which is why "grubbing for roots" takes on so much importance.

Degradation proceeds in other ways, too. Ladku was a tenant-cultivator on about 25 hectares of land belonging to a rich non-farmer from a town far away. There were about 1,500 trees on the land of which Ladku cut down only four trees a year for firewood and sale, and lopped off the small branches of others for manure. He freely permitted other Warlis to collect branches and leaves from his land.

Ladku fought for many years to get the land transferred to his own name under the land-to-the-tiller laws and finally succeeded in 1986. However, the laws allow the previous owner to cut down all the trees on the land except for a few fruit species. Most of the cut trees sent out coppice shoots immediately but now these are insufficient to cater to the needs of Ladku and the others. The result is that all the shoots are being cut and there is little chance for regeneration at all. The situation will deteriorate further every year since some of the trees will be unable to send out further shoots.

Not all the Adivasis are innocent. Vithal from Babhan pada has turned into a timber merchant. He visits forested areas, hires some local labourers, cuts the trees and sends them to Vapi, a neighbouring industrial town. It's a flourishing business. Vithal says he gets Rs 200 per ton. "Do you get enough trees?" "Oh yes! There are lots. You can easily earn Rs 600 per week. That much wood you can get, but who has the time and means to co-ordinate everything?" "But this way, the jungle will be destroyed and one day there will be nothing left." "But if I don't cut these trees, some other fellow, a trader, will cut them."

Some years ago a huge dairy scheme was implemented in Dapchari, supposedly for the benefit of the Warlis. More than 1,500 hectares of Warli land was requisitioned and subdivided into small plots. Each Warli was offered a plot and two crossbred cows on a loan of Rs 36,000. A further large area of Warli land and jungles was acquired for a dam and lake to supply water to the scheme.

Very few Warlis took advantage of the scheme and those who did not were offered alternative land which turned out to be so useless that they refused that too. Those who did take up a dairy plot soon found that they were unable to make a profit in spite of the advisory, veterinary, fodder and other services provided by the scheme, because of the usual problems with crossbreeds. The authorities claimed that the failures were on account of the inability of "backward tribals" to learn modern techniques. This excuse was employed to give the plots to non-Adivasi, "educated" persons from as far away as Punjab. But these too have been unable to make profits since the fault lies in the Western technology itself.

A few Warlis were given jobs in the scheme but they were paid so badly that the Communist Party organized a strike. After a long struggle they won substantial pay increases with retrospective effect but the Party took contributions of Rs 1,000 from each worker.

The canal from the dam to the project area leaks considerably with the result that the water table is rising in surrounding areas. In some places it is just 50 centimetres below ground level and soon the Warlis will not be able to grow anything in the area. The numerous pools of stagnant water provide ideal breeding grounds for mosquitoes.

Along the Bombay–Ahmedabad highway as you come to the Saoroli crossing, you see a board announcing the Shrinath Industrial Estate. Some factories are already operating, others are coming up fast. The unskilled wage rate is Rs 10 per day. One day, as we were going along a road near the estate, Prakash Pachalkar told us that the whole area covered by the estate was formerly a jungle. Now all the trees have been cut down and only a few monsoon annuals are to be seen. The effluents from the factories are killing off these too, while the pollution of ground water has affected the neighbouring crops. Only a few Warlis have obtained jobs in the factories since industrialists prefer to employ already-trained labour from the cities.

The conversion of the commons under social forestry schemes is rapidly restricting access to these, too. In Moho, the forest department decided to grow mulberry trees on four hectares of Warli pasture land. While a few may have earned extra income from rearing silkworms, all in the village would suffer because there would not be enough pasture left for their cattle. The people managed to stop that project, but in the following year, the whole of the grazing grounds were planted with *nilgiri* in spite of the protests of the villagers. The forest officers said that they had too many *nilgiri* seedlings which had grown too big and since no one wanted them, they dumped them on this village. The villages countered by encouraging their cattle to trample on and kill the plants.

The government's Employment Guarantee Scheme is supposed to ensure that any unemployed person gets work on schemes of public utility. Not all who need work accept such labour because they find they have to travel too far and it does not leave them time for their domestic and agricultural work. Many of the schemes serve only the rich farmers: on the land of Bhanushali, a moneylender, a dam was built as a watershed development project.

Other Warlis prefer to migrate in the dry season to towns and cities. There they work at brick manufacture, dredging sand and other operations which are carried out only in this season. Often these are given advances by those who run these operations so that in effect they become bonded labourers.

The mainstream now lures the Warlis into the cash economy by its projection of consumer items as symbols of civilization. Those brought back as proof of entry into the mainstream are pitiful: factory shoes, the latest style in shirts, tinny transistor radios. Warlis have always used several plant materials as detergents, for instance, the fruit of *ritha* and *shikakai*, and the leaves of *erandi*.

However, companies such as Hindusthan Lever, avid to capture the large potential rural market, vigorously sell their soaps in every village.

The institution of moneylending still thrives, at exorbitant rates of interest. Bhanushali owns large areas of land and amasses hundreds of tons of grain as interest soon after the harvest. Bhanushali cannot store grain for more than a year, so accumulation of perishables is impossible, but cash can be banked. Some years ago, a leftist organization drove out the moneylenders from a number of villages. But because common resources were not restored, some of the Warlis themselves became moneylenders.

The synergistic and cumulative effect of all the modes of impoverishment has brought down many of the Warlis to below subsistence level. Shevu explained: "We eat in a room without windows because we do not want anyone to see what we are eating. There may be times when we have to go hungry."

The benefits of industrialization are claimed to have been mainly responsible for the rise in life-span in the West. But that has been brought about by a corresponding decrease in the life-spans of the poor. A transfer of wealth is invariably followed by a transfer of health. All the processes of material appropriation result in a lowering of the Warlis' life-span.

RECOVERY

As Kings of the Forest, Warlis have the right to use it. This right cannot be abrogated by any laws that invaders make, even if they claim that they are in the national interest.

The preservation of the commons requires an appreciation of our total dependence on them. The mainstream needs to absorb the cultural values of the Adivasis, not the other way around. We need to have faith in the validity of the principles on which their system is based. This requires a radical change in ourselves, in our way of thinking and in our values. Indigenous knowledge is often not seen as valuable in itself but merely as a means to introduce Western concepts of development.

Enforcing conformity with mainstream beliefs through the educational, health, agricultural and other sub-systems is a certain method of destroying sustainable knowledge. Modernization, if necessary at all, should be an adaptation of good traditional principles and methods to the problems of today, with inputs from new science and technology used only where they are not self-destructive.

The privatization of property, the enclosure of the commons and the growth of the cash economy are claimed to be advances of civilization. But these processes also permit and promote oppression, and are essential to the process of enrichment of a few at the cost of the impoverishment of the majority. True justice may not be attainable until the processes that make people oppressed are removed or reduced.

An old Indian proverb asks, "who sees the peacock dancing in the jungle?" Few do, but the beautiful dance is an integral part of the survival of the species.

There are many elegant but unseen examples of lifestyles that are based on principles of true sustainability: the ability to survive for millennia without degrading the environment. They enhance the lives of those who practise them, and offer instruction to those who care to listen.

NOTES AND REFERENCES

1. David Hardiman, *The Coming of The Devi: Adivasi Assertion In Western India* (Delhi: Oxford University Press, 1987).
2. Most of the information on British land policy, taxes, *daru* and *tadi* is taken from the book mentioned in note 10.
3. S. R. Paranjpe, "Cocoanut Cultivation in the Konkan", *Poona Agricultural College Magazine*, January 1912.
4. D. D. Kosambi, *Exasperating Essays*, (Pune: Nene, 1957).
5. Quoted by G. F. Keatings, Director of Agriculture, Bombay, in "Agricultural Progress in Western India", *Poona Agricultural College Magazine*, July 1913.
6. Yashodhara Dalmia, *The Warlis: Tribal Paintings and Legends*, (Bombay: Chemold Publications and Arts).
7. Ibid.

3. Red Ink in the "Blueprint for a Green Economy"

It is interesting to observe that the West, whose system of development did so much to destroy sustainable practice in India, has recently been converted to the virtues of "sustainability". Furthermore, the western system of development is widely assumed to be necessary and sufficient for promoting social justice. There are several reasons why this assumption can be questioned, but it is the environmental issue which is now raising the most doubts.

Evidence for the existence of environmental crises has been accumulating for decades. The prime threats at the moment appear to be global warming and the disappearing ozone layer. The near future will probably add the widespread pollution by toxic chemicals and the release of genetically modified organisms into the environment. Any one of these could be catastrophic: the threat to animal and plant life on earth has been acknowledged by even the most conservative of the world's leaders.

Western and westernized politicians, economists, scientists, and industrialists, until recently, ignored the warning signs. But one of the crises – global warming – seems to be arousing them from their mental hibernation.

Today, scientists no longer discuss whether the earth will warm up; they are busily computing how hot it will get, and speculating what the consequences will be. Politicians and economists are fast turning a chameleon green, proclaiming new policies which, they claim, are environmentally benign. One might have expected those in power to attack the source of the problem. Instead, they are making desperate attempts to justify the continuance of a way of life which has itself generated the present crises.

The point of contention is "sustainable development". It is

claimed that sustainable development is compatible with a high standard of living, based on extensive consumption of goods and services. It is even asserted that a rising income is indispensable for sustainability.

These claims are partly based on the report written by the World Commission on Environment and Development called *Our Common Future* (often called the Brundtland Report, after the Norwegian Prime Minister, which is often quoted with a reverence worthy of an Old Testament revelation: "Our report, *Our Common Future*, is not a prediction of ever increasing environmental decay, poverty, and hardship in an ever more polluted world among ever decreasing resources. We see, instead, the possibility for a new era of economic growth".[1] This report then goes on to give frightening details of environmental decay and its consequences; it does not provide a guide to the promised land of uncontaminated milk and money.

A new report, *Blueprint For A Green Economy*, by David Pearce and his colleagues, claims to furnish a reliable route to the goal promised by Brundtland.[2] This could become the New Testament of all right-thinking people.

The basic principles set forth in the report are sound. Most of the methods suggested are unexceptionable, even exemplary. However, the recommendations are diluted with so many qualifications and loopholes that what they claim to be doing is radically falsified. Even if their suggestions were fully, logically and rigorously implemented, these would still be powerless as a prescription for growth with environmental enhancement. The contradictions remain unresolved.

THE PEARCE REPORT

The Definition

The authors state:

> Sustainable development involves a substantially increased emphasis on the value of natural . . . and cultural environments . . . . Sustainable development places emphasis on providing for the needs of the least advantaged in society ("intragenerational equity"), and on a

fair treatment of future generations ("intergenerational equity")...
The well-being of the most disadvantaged in society must also be given greater "weight" in a developing society: if average well-being advances at the cost of a worsening of the position of the most disadvantaged it seems reasonable to say that such a society is not developing. . . . It is perfectly possible for a single nation to secure a sustainable development path . . . but at the cost of non-sustainability in another country.

Let's get this clear. "Intragenerational" implies all persons now living, in every corner of the earth. "Intergenerational" implies all future generations – our own children and theirs for an indefinite number of generations. Equity requires an undegraded environment, one of the requirements for social justice. With these definitions Greens can have little quarrel.

It is the Report's qualifications of these definitions that make it untenable:

> Development is some set of desirable goals or objectives for society. These goals undoubtedly include the basic aim to secure a rising level of real income per capita – what is traditionally regarded as the "standard of living". . . . Sustainable development involves devising a social and economic system which ensures that these goals are sustained, i.e. that real incomes rise, that educational standards increase, that the health of the nation improves, that the general quality of life is advanced.

Confusion is sown by the coupling of the terms "sustainable" and "development". The authors use the word development as understood by the industrialized West and then assert that this has to be sustained. Sustainable development then is reduced to mean sustaining the Western economic system. But they have to prove that economic growth is compatible with environmental quality and equity; they cannot just postulate it as if it were a divine revelation.

They present a graph (reproduced on p. 65) supposedly showing the variation of real incomes with environmental quality. A straight line, C–C, through the origin, depicts real incomes increasing linearly with environmental quality. We can assume that incomes are related to expenditures on manufactured products and services. As will be seen later, nearly all these have detrimental environmental effects. There is no evidence to indicate that a line such as C–C can exist. The TO–TO curve is a more reliable indicator of the impact of expenditure on environmental quality.

The authors grant that:

> The "growth versus environmental" debate is clearly still a real one in the sustainable development context. . . . For the moment we observe that growth does not *necessarily* involve environmental degradation. The view that it does is based on perfectly respectable intellectual foundations. . . . It is clear that materials get used up in all economic processes. If the materials are in finite supply then, by definition, the faster is growth the faster is their depletion. That raises the prospect of "running out" of resources. . . . There are limits to recycling. . . . We cannot recycle energy at all. Once used it is dissipated. . . . It is another type of resource that is in scarce supply – the resource of the natural environment as repository for all the waste products associated with materials and energy use.

The "respectable foundations" are provable. Growth, as well as the present levels of consumption, "necessarily involves environmental degradation".

All materials, renewable and non-renewable, are in finite supply. The energy component in all manufactured goods and services, in production, processing and transport, renders them intrinsically polluting and incapable of being recycled.

The major sources of industrial energy are highly polluting. There is a limit to which the carbon dioxide produced by thermal power stations can be fixed by trees. Once a forest has reached its climax stage, it may not fix any more carbon dioxide. Such trees should not be used for fuel, because the carbon should be permanently locked up in them. Planting forests would, at most, delay environmental degradation by a few years. Nuclear energy generates radioactive wastes which cannot be de-activated by any means, and their storage will degrade the environment for thousands of years.

If renewable sources such as solar, wind and hydro power were to replace all thermal and nuclear power, they too would cause heavy environmental damage. Big dams displace people and submerge forests and fertile lands, large wind generators produce noise pollution, tidal power barrages destroy sensitive coastal

CHARACTERIZATION OF TWO APPROACHES TO GROWTH AND ENVIRONMENT

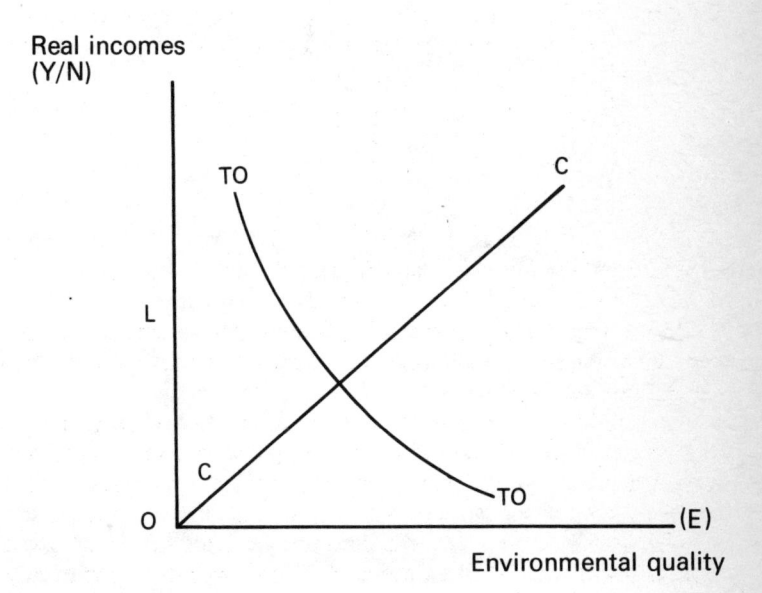

Real incomes
(Y/N)

TO

C

L

C

TO

O ──────────────────── (E)
Environmental quality

In this simplified diagram, society has to choose where to locate itself in terms of the amount of economic growth (an increase in real income per capita, Y/N) and the amount of environmental quality (E) that it wants. If growth can only ever be achieved at the expense of environmental quality, then society has to choose a point on TO–TO. If, on the other hand, growth and environment are wholly compatible, it chooses a point on C–C. A "limit to growth" could be typified by a point such as L where hypothetically, growth brings about zero environmental quality – i.e. a "doomsday" solution. (Reproduced from *Blueprint for a Green Economy*.)

environments. If biomass energy is used, the energy extracted must not produce more carbon dioxide than that which is being fixed by plants, if it is not to be a pollutant.

Additional waste products which are overloading the environment are other greenhouse gases; CFCs and several more ozone-depleting gases accumulating in the atmosphere; DDT; PCBs and thousands of other toxic chemicals which have permeated the environment from pole to pole; synthetic fertilizers and pesticides which are contaminating water and soil; and many other environmentally-malign products.

Every second that an electric light or appliance is left on, every centimetre that a vehicle is driven, almost every product and service used, adds to environmental degradation.

Economic growth can be made compatible with environmental enhancement only if the emission of pollution is less than that which can be degraded by the natural environment. In order that resources be conserved, all articles must be manufactured so as to be fully recyclable. Further, they must be manufactured, transported, used and recycled with energy from renewable sources only. On a second-order level, renewable energy generators and transmission equipment would need to be manufactured on the same basis.

Until this is accomplished, a high standard of living implies a high level of pollution, resource consumption and violence to the earth. It is, literally, costing the earth. Economic growth or even the maintenance of a high standard of living, is irreconcilable with a thriving environment. This is why equity is not possible within existing industrialism. And the higher the consumption, the higher the inequity actively produced, the higher the rate of social exploitation. Every use, in the words of *Our Common Future*, "compromises the ability of future generations to meet their own needs".

The authors of the Pearce Report continue: "The essential point however is that the relationship between materials, energy flows and environmental waste sink capacity to economic growth is not immutable. . . . If these ratios can be systematically reduced, then growth *with* environmental quality is feasible". But, at most, the modifications suggested will to a minute degree reduce the rate of increase of damage to the environment, even while its absolute value keeps on soaring. By ignoring the most important factors in their calculations, the authors achieve the apparent confirmation of the compatibility of economic growth with environmental quality.

In the final analysis, it is selfishness that prevails: "Very simply, given limited resources, the rational thing to do is to choose between our preferences in an effort to get the most satisfaction – or 'welfare', to use the economist's term – we can". The authors, however, do realize that welfare cannot be the sole criterion in a deteriorating environment: "Some care needs to be exercised, then, that the use of social objectives such as gains in welfare does not dictate or support policies which are inconsistent with the ecological preconditions for existence or, at least, some minimal quality of life".

Moreover, the authors dilute their definition further: "Achieving economic development without sacrificing an acceptable rate of economic growth may be said to define the problem of sustainable development". So, sustainable development is ultimately to be determined by an acceptable rate of economic growth, not by sustainability! As the authors themselves state: "There is some truth in the criticism that [sustainable development] has come to mean whatever suits the particular advocacy of the individual concerned".

To distinguish between meanings, the term sustainable development will be used as the report uses it here and the term sustainability will indicate the rigorous, holistic definition, which puts the environment before economics.

Bequeathing equivalent capital

The authors' prescription for achieving sustainable development is "to leave to future generations a *wealth* inheritance – a stock of knowledge and understanding, a stock of technology, a stock of man-made capital *and* a stock of environmental assets – no less than that inherited by the current generation".

It is normal accounting practice to take into consideration liabilities as well as assets, but the authors do not mention them, even though these have crucial consequences for intergenerational as well as intragenerational equity.

Intergenerational liabilities

"Pareto optimality [a situation where it is not possible to improve one person's welfare without damaging another person's welfare] is of course consistent with making the future better off so long as the present is not made worse off. This last proposition should serve to remind us that the future has an obligation to the present just as much as the other way round."

The future will have an obligation to the present if, and only if, the people of the future are bequeathed net assets, not liabilities. Moreover, future generations have an inalienable right to an unspoilt, unexploited environment, whereas the present generation has a right only to what can be extracted without diminishing the natural capital.

Further, intergenerational equity must take account of the past as well as the future. The present stock of capital wealth in the West has been accumulated by exploiting the Two-Thirds World, through colonialism and neo-colonialism, and by mining natural capital resources on a massive scale. The present balance sheet should reflect the liabilities which have accrued during earlier centuries.

The first charge on the present capital wealth, then, is its restitution to the descendants (the ex-colonized as well as indigenous peoples in still-colonized countries) of those who have been deprived of their economic capital and natural wealth.

There is also the huge "brought-forward" global pollution and degradation which, unless corrected, will be passed on too. The environment must be restored to its uncontaminated state, which means claims on the present wealth, since it has been partly accumulated by polluting it. It is necessary not merely to pass on environmental assets equivalent to those received by the present generation, but to bequeath much more.

There are liabilities involved even in the stock of knowledge and technology. Cultural colonialism has led to the destruction of knowledge and skills which made Two-Thirds World traditional systems sustainable.

One can claim that it is unfair that the present generation bears the costs for the benefits which accrued to the people in the past. But the present generation is also benefiting (in the authors' meaning of the term) from the past accumulation of assets.

Current liabilities include further resource depletion and pollution which are being progressively added to in direct proportion to consumption. The greater the environmental damage already inflicted, the greater the harm each action contributes. The higher the standard of living, the higher the rate of increase of the burden left to the future. As economies keep on growing, the rate is also rising, perhaps exponentially.

If these liabilities are not discharged, there will be an enormous net liability bequeathed to future generations. The existence of such liabilities is not even acknowledged by the authors. Instead, they take present conditions as an initial benevolent or neutral state, and go on from there to cost future interactions only. Those who are so reluctant to consider the remission of monetary debt in the Two-Thirds World are happy to write off environment debt that is incalculable.

Even if it is assumed that the slate is clean now, the authors themselves pose further problems:

> What is being said is that we can meet our obligations to be fair to the next generation by leaving them an inheritance of wealth no less than we inherited. Moreover, so long as each single generation does this, no single generation has to worry about generations far into the future. Each generation "looks after" the one that follows. On the face of it, this solves one otherwise intractable problem of deciding how far into the future one needs to look in order to decide how "sustainable" current development activity is. In practice, problems remain because the effects of some current activities will spill over to many generations to come: the storage of some radioactive waste, for example, and the loss of biodiversity. . . . The *theoretical* approach in economics is to analyse the problem in terms of "infinite time horizons" – i.e. not to set any limit in time to the analysis. Again, this convention permits powerful techniques of analysis to be used. We do not adopt it here because the aim is to "translate", as far as possible, some of the findings of this economic analysis into politically more operational rules.

Consistency requires that these effects be taken into account and not "translated" to exclude anything that makes politicians uneasy. But the authors, while admitting that their formulae will still burden future generations with liabilities, quietly forget all about the satisfaction of intergenerational equity.

The equivalence doctrine

To compensate for environmental degradation, the authors suggest replacing natural resources with artificial ones. "The implicit assumption in the first definition of sustainable development is that man-made and natural capital are *substitutes*. That is, so long as the overall *aggregate* of natural and man-made capital does not decline between one generation and the next, the stock of natural assets can decline because the growth of man-made capital will compensate for it." Future generations are expected to be grateful if the bountiful, life-giving yield of a natural forest, priced merely at its timber and tourism value, is replaced by contaminated concrete of equivalent economic value. It is a more than linguistic irony that the living richness and variety of the jungle can be seen as indistinguishable from what is familiarly called "the concrete jungle".

However, a little later on the authors discredit their own dogma:

> The alternative approach is to focus on natural capital assets and suggest that they should not decline through time. *Each generation should inherit at least a similar natural environment* . . . [original emphasis]. It does not imply that the two forms of capital are perfectly substitutable. . . . Essential functions of the environment, such as complex life-support systems, biological diversity, aesthetic functions, micro-climatic conditions and so forth, have yet to be replicated by man-made capital, or can only be substituted at an unacceptably high cost. . . . In general, there is no easy interpretation to the idea of a constant capital stock.

But, having admitted that the two types of capital are not equivalent, they still state: "It can be shown that such a 'constant capital' bequest is consistent with the concept of intergenerational equity. . . . It is the aggregate quantity that matters and there is considerable scope for substituting man-made wealth for the natural environmental assets".

RED INK IN THE "BLUEPRINT FOR A GREEN ECONOMY" 71

If there is no clear idea what constant capital stock is, how can they claim that equivalence is consistent with intergenerational equity?

They construct a further codicil through which they can escape their own suggested bequests:

> *Future generations should be compensated for reductions in the endowments of resources brought about by the present generations.* . . . Hartwick showed that a society with an exhaustible resource, such as oil, could enjoy a constant stream of consumption over time provided it invested all the "rents" from the exhaustible resource. (A "rent" is the difference between the price obtained for the resource and its costs of extraction.) What Solow shows is that the Hartwick rule is formally equivalent to *holding the overall capital stock constant* [original emphasis].

We may well enquire in what cosmic legislature such "laws" and "rules" were enacted, and over what nebulous regions their writ runs.

Can "rents", however high and judiciously invested, ever cover the cost of replacing non-renewable resources? At the present time, these rents are not even being used for developing alternative resources, even for mineral oil.

Moreover, as the authors state: "If we extend the bequest motive to future generations in general, as many environmentalists would urge us to, we face the difficulty of not knowing their preferences". One might have thought it possible to predict that the first preference of all future generations would be survival.

> In a society which is based on letting people's preferences count, as market-based economies are, it is not logical to accept the role those preferences play in the allocation of goods within society *now*, while rejecting the preferences that people have for the present over the future. Respecting time preference is just as much a feature of "consumer sovereignty" as respecting people's rights to buy and sell what they choose in the market place.

Here, the authors reveal the constraints on their project: the market-based economy is sacrosanct. No neater apology for business-as-usual could be devised by the most ardent proponent of unbridled industrialism. Consumer preferences – in truth, manufacturers' preferences – override environmental or equity considerations.

Intragenerational liabilities

"Pareto optimality is achieved when it is not possible to make one person better off (improve his or her 'welfare') without making another person worse off." This seemingly fair statement hides the fact that the present state is highly inequitable. As Amartya Sen says: "A state can be Pareto optimal with some people in extreme misery and others rolling in luxury, so long as the miserable cannot be made better off without cutting into the luxury of the rich."[4] Pareto optimality effectively freezes the injustice in the system. It renders the rich inviolable.

The authors suggest: "A constant or rising natural resource stock is most likely to serve the goal of intragenerational equity when the productivity of ecosystems is essential to the livelihoods of the poor. In such instances we are really talking about preserving *sustainable livelihoods.*" They do not seem to be aware that most of the ecosystems necessary for preserving livelihoods have already been badly damaged, if not wholly destroyed, by historic and current exploitation which has sustained western unsustainability.

The authors state:

> A nation which could be regarded as importing sustainability should seek to compensate the exporting nation for its loss. Essentially, the exporting nation risks non-sustainability for the benefit of the importing nation. . . . This idea takes us beyond the remit of the current report, but effectively it points us towards the importance of *Foreign aid for sustainable development* [original emphasis].

By importing timber from Indonesia, for instance, the UK is importing sustainability and exporting unsustainability to Indonesia. Consistency demands that the timber should be valued taking all the environmental effects of rainforest destruction into account. The UK should then PAY for Indonesian timber at this price. This would reduce neo-colonial exploitation, price the product appropriately, give Indonesia more funds so that it needs to cut down less forest, prevent further environmental damage and perhaps correct some of that done earlier. But if such pricing were rigorously followed, it would swiftly lower the revered "standard of living". This is why graciously-given charitable aid is offered instead of economic and environmental justice.

Irreversibilities

The authors warn that much damage to the environment is uncorrectable and permanent:

> A particular extreme instance of such rising costs occurs when the damage done by delay is irreversible. . . . The damage done, for example, through the contribution to climate warming is irreversible. . . . This suggests a variation on the "constant wealth" idea, namely that it should be pursued subject to the avoidance, as far as feasible, of irreversibilities. . . . To maintain diversity it is essential to avoid irreversible choices.

As we have seen above, practically every consumable adds to climate warming, but no recommendations are made to cost these now. The other irreversibilities, which the authors refer to, are the extinctions of plant and animal species. Such irreversible damage is daily being aggravated.

Moreover, all our interventions are irreversible. There is a story in which a man travelled in time millions of years into the past. While there, he accidentally killed a butterfly. Returning to his present, he found that the benevolent state in which he had lived was now an atrocious tyranny. Every intervention of ours changes history. Mathematical theories of chaos now confirm that small interventions can have large consequences.

Further, all our actions involve the use of energy which leads to an increase of disorder, of entropy, according to the Second Law of Thermodynamics. When we use nuclear fuels we are using the energy stored by stellar processes billions of years ago. When we use fossil fuels we are using solar energy stored over millions of years. There is no way in which human beings can re-store this energy after use. When we use them, we are irreversibly consuming a patrimony to which future generations have an equal right.

Plants alone naturally increase order on the earth, with the help of external solar energy, within our lifetimes. The destruction of forests then amounts to reducing the order produced on earth, and the human-made wilderness which replaces them adds much more disorder. We can justifiably use only what we ourselves replant. Other renewable sources of energy should also be used minimally, since we are ignorant of the long-term effects they produce.

"Since knowledge is rarely lost for ever, irreversibility involving man-mad capital is rare – discontinued machines can be reconstructed, structures rebuilt, technology recreated, and so on." But most of the knowledge of the West is knowledge of how to produce irreversibilities. The very notion of progress itself, so crucial to Western ideas of development implies a one-way process, that is, an irreversible process.

In any case, even deconstructing – for instance – the Narmada dams, or other mega-projects, cannot restore the patrimony of the displaced Adivasis or the life lost in submerged jungles.

Pricing the environment

"Economics throws light on the meaning of sustainable development because the environment and the economy *necessarily interact.*" It is this rather belated discovery and the inability to see clearly all its implications, that is the root of the crises today.

"In benefit estimation money is used as a measuring rod, a way of measuring preferences. . . . As long as we do not forget that there will be some immeasurable gains and losses, the measurement of gains and losses in money terms will turn out to be revealing and, we shall argue, supportive of environmental values and environmental policy." The environment is thus still recognized for its economic value only; it is not accepted that a sound and wholesome environment is essential for our survival.

The authors admit that there are factors that are priceless: "There are, however, two interpretations that might be placed on the idea that something is priceless. The first is that priceless objects are of infinite money value. . . . The second . . . says that there are some things in life which simply cannot be valued in money terms – there is somehow a compartment of our thinking that refuses to place money values on, say human life". Such delicacy of feeling probably comes from the fact that if human lives were to be priced by Western economics, the differential between that of a

Westerner and a child dying of malnutrition in the Two-Thirds World might be embarrassingly divergent. By their curious logic (which is also that of Western economics) what cannot be valued because it is immensely valuable, is omitted from the balance sheets.

Even the items that are valued are assessed in a very restricted fashion: "Preserving a wetlands area, for example, may well be at the cost of agricultural output had the land been drained". This is a simplistic version of the contribution of wetlands to the biosphere.

"To avoid the risk of misunderstanding: what is being argued here is that, *given* an environmental quality objective, the aim of society should be to achieve that standard at minimum cost." It is difficult to quarrel with this, even though one might object to the way in which the use of language devitalizes the living environment. However, the "environmental quality objective" must not be itself determined by "acceptable" costs. It should take full account of the criterion of equity that the authors themselves advocate. And the authors adapt the objective to continuing economic growth.

In pricing the environment, the authors say: "But sustainable development attempts to shift the focus to . . . ensuring that any trade-off decision reflects the full value of the environment". Good costing-practice would mean asking the person who directly benefits to bear all the costs, too. Doing so would conspicuously expose the injustice in a hyperconsumptive society. Hyperconsumption is possible only because consumers bequeath costs while enjoying the benefits. The user of energy has to bear only the diffuse costs of global warming, but within the next few decades millions could be displaced by a rising sea level. This is the case with most industrial commodities. The user gets the benefits of the article, while the pollution intrinsic in its manufacture, use and disposal devolves on others.

If it is to reflect the full value of the environment, the price of an item would include the cost of manufacturing it by non-polluting methods, or the cost of removing the pollutants emitted in its production. Merely trying to eliminate the main pollutants – the greenhouse gases and those that damage the ozone layer – must necessarily raise the cost of goods and services to such a degree that few could afford them.

There are, also, what the authors call existence values: "Some empirical attempts to measure existence values tend to relate to endangered species and unique scenic views". These are expressed by equations such as:

> Intrinsic value = existence value, and Total economic value = actual use value + option value + existence value.

But no example of the application of these equations is given anywhere in the book. This is understandable: because the full value of the environment includes non-monetary factors, the very process of trying to determine a total economic value is doomed to fail.

Uncertainty

In concluding the chapter on "Valuing the Environment", the authors grant that "it is not essential to be persuaded that the monetary valuations illustrated in this chapter are 'accurate'". All that their methods emphasize is that "environmental services are *not* free", that valuing the environment makes decisions more rational, and so on. A little further on, they present pages of "Difficulties in the application of estimation techniques". Many of the estimated errors are large, two of them as much as plus or minus 60 per cent. With such uncertainty, the validity of their methods becomes doubtful even on their own terms.

Perhaps realizing that any attempt at correct costing must show that most projects would be harmful, the authors open up another escape hatch:

> Essentially, the economic efficiency objective is modified to mean that all projects yielding net benefits should be undertaken subject to the requirement that environmental damage (i.e. natural capital depreciation) should be zero or negative. However, applied at the level of *each project* such a requirement would be stultifying. Few projects would be feasible. At the *programme* level, however, the interpretation is more interesting. It amounts to saying that, netted out across a set of projects (programme), the *sum* of individual damages should be zero or negative.

Such manipulation would, no doubt, authorize writing off carbon-dioxide emission in the West against forest growth in the Two-Thirds World.

Incremental environmental benefits and sustainability

Taking environmental costs into account obviously helps. But if all environmental costs are not taken into account, incremental changes may show deceptive benefits. For instance, the elimination of lead in fuels is claimed to give a net benefit of $6.7 billion through greater health and welfare. But the continued use of lead-free fuels keeps on adding to global warming, acid rain and other problems. The total effect on present as well as future generations will remain a large liability.

The possibility of incremental improvements cannot, therefore, be used to support the argument that economic growth with sustainable development is possible.

Inertial guidance system

The authors concede:

> If we do not know an outcome it is hardly consistent with rational behaviour to act *as if* the outcome will be a good one. If environmental risks have the potential for large negative payoffs then risk-aversion dictates that we protect natural environments, at the very least until our understanding of how they function in terms of life-support grows.

In the case of global heating the authors admit that "the outcome if the *worst* happens is clearly catastrophic". Unless the authors do not consider catastrophe a "large negative payoff", "risk-aversion dictates" that we must act now. Yet they present economic arguments for delaying action. "It may therefore make sense not to act precipitately about an environmental problem such as climate change". . . "The implicit condition here is that the delay is accompanied by further research aimed at such cost-effective solutions". . . "In fact, however, 'doing nothing', or undertaking a 'research-oriented-do-nothing' policy may be justified if it can be shown that (a) action now would be very expensive, or (b) more cost-efficient action might result from delaying the intervention policy."

Because scientists cannot say precisely how much the temperature will rise, it does not mean that no action has to be taken to prevent any rise at all. There is no way of predicting whether a lower-cost solution may be available through further research unless economists have successfully tamed an oracle. Moreover, thermodynamics says that such solutions cannot be found in many instances. And there are the uncorrectable cases: how does one lower the sea level after large areas of Bangladesh and the Maldives are submerged? How does one resurrect a single person killed by pollution?

A further reason for postponing action is discovered in the case of transborder pollution: "No one country acting alone can do much to prevent or contain these impacts: only a coalition of governments worldwide can do this". The effect of this is to offer individual countries a licence to do nothing beyond issuing pious declarations.

Although the US accounts for less than 5 per cent of the global population, it generates about 24 per cent of carbon dioxide and 25 per cent of nitrogen oxides. The Soviet Union accounts for 19 per cent of the carbon dioxide, China 9 per cent, and Britain 3 per cent. The industrialized countries currently consume about 80 per cent of the ozone-depleting chemicals. The US and the European Community together consume 60 per cent, Japan 12 per cent, the Soviet Union 9 per cent. It is evident that these countries acting alone could make a substantial contribution to "prevent and contain impacts".

At a recent ministerial meeting in the Netherlands, the US, USSR, UK and Japan refused to agree to stabilize carbon-dioxide emissions by the year 2000, although 60 other nations consented to do so.[4] That conveniently leaves all of them free to add to the "ever increasing environmental decay, poverty, and hardship in an ever more polluted world among ever decreasing resources", which Brundtland bluntly predicted would not happen. It is clear that the necessities of economic systems in the West readily vanquish archaic notions of Christian or indeed any other morality.

This does not mean that Two-Thirds World nations should not act on their own. But a clear and valid distinction needs to be made between the use of resources for – and the pollution produced by – processes essential for survival, such as cooking fires, and those that are superfluous such as private, air-conditioned cars, even though economists may see no difference in their environmental costs.

Postponing action dumps a liability on our descendants which in all probability they will be unable to handle. Burdening future generations with research is itself transferring liabilities to them. The authors do conclude that "a society committed to sustainable development will shift the focus of its environmental policy to an anticipatory stance". But there is no emphasis on the urgency of this task.

Scientists' ignorance

In order that the prescription be fully implemented, it is essential that we have knowledge of how all our interventions affect the environment. But such knowledge does not exist. After discussing the uncertainties about how and where environmental impacts will manifest themselves, the authors point out: "To these kinds of uncertainties must be added the scientific uncertainty about how ecosystems function – consider the fact that we are some way away from understanding how global carbon cycles work in detail and hence how climate change will impact regionally within the world".

Scientists are often unable to foresee the extended consequences of their actions. To give one example, they are still discovering the harmful effects to introducing synthetic fertilizers which have been strongly promoted for many decades. At the time they were introduced, no scientist knew that replacing organic manures with artificials would:

- reduce the contribution of rhizobia and other nitrogen-fixing organisms;
- kill off organisms that keep the soil healthy;
- encourage pests to proliferate;
- contribute to soil erosion on a terrifying scale;
- pollute water and cause cancers;
- release nitrogen oxides to add to smog, ozone-layer depletion and greenhouse heating;
- decrease the amount of methane metabolised by soil organisms and so make another contribution to climatic change; *and*

- taking all these into account, reduce agricultural outputs.

No one can put a reasonable figure even now on the costs of using synthetic fertilizers, since they may cause still unknown problems.

Scientists can also make errors in interpreting data. When the ozone hole was first detected by satellite-borne instruments, the data were rejected as instrumental error since there was no theory to explain them.

An editorial in the *New Scientist* grants that science is to blame for the environmental problems of today. "Even the most die-hard techno-enthusiast must admit that many of the environmental problems which have given rise to the need for a Green movement are the unanticipated side effects of a massive social investment in science-based technologies. There was a time when chemicals such as CFCs or DDT were widely seen as both marvels of science and as major contributions to social progress."[5] If science is to blame for causing the problems, it is rather unlikely that that same science driven by the same motives, funded by the same sources, and with the same reductionist structure, can solve those problems.

The few solutions to environmental problems that scientists have come up with border on the bizarre. A scientist claims that the destruction of the ozone layer could be slowed down by beaming radio waves into the stratosphere.[6] No calculations have been given by him of the amount of additional pollution that will be created by the power used to produce the waves.

The USSR claims the ozone layer can be repaired by using planes to introduce ozone-generating substances at altitudes of 22–24 kilometres.[7] Nothing is said about how much more ozone will be destroyed by the pollution produced by the airplanes themselves.

Some scientists even say that global warming is a good thing. Mikhail Budyko, the Soviet Union's leading climatologist, was the first to predict that the greenhouse effect could warm our planet. Now he claims that the warming will be beneficial because temperature rise in the USSR will result in increasing agricultural production.[8]

The most dangerous suggestions are that science will be able to cope with all the possible changes that could occur.

Biotechnologists, for instance, claim that they will be able to modify plants and animals so that they can be adapted to climatic change.

One scientist whose theories suit the establishment most comfortably is James Lovelock, with his Gaia hypothesis. An article states: "He believes that Gaia, the living planet, has built-in life support systems which can soak up pollution and act as a planetary thermostat".[9] Lovelock claims that acid rain, radioactivity, toxic chemicals like PCBs and the ozone hole are minor issues. The only problem that worries him is disruption to the carbon cycle by deforestation. Lovelock states that technology must be used wisely to separate us from nature, even from Gaia itself.

The West isolates itself from the real environment by imposing a technosphere upon the biosphere, which is then mistaken for reality itself. The technosphere has become a surrogate life-support system that obscures our true dependency on the natural world, the sole giver of life.

The industrial solution: green consumerism

Resort is also taken to the "magic of the marketplace". "If consumers change their tastes in favour of less polluting products and against more polluting forces, market forces are supposed to lead to a change in the 'pollution content' of final products and services. This is the 'green consumerism' argument."

But changing to less polluting products is not enough to provide sustainability. No industrialized product is fully green and making slight improvements will not prevent drastic degradation. Further, "for green consumerism to be effective consumers must be informed about the pollution profile of the products they buy". It is rather unlikely that manufacturers will loudly advertise all the harmful effects of their products – even those of which they are aware.

Moreover, it is the free-market economy that has led to the grave environmental crises of today. One cannot see how the same mechanisms which have produced the disease can cure it. Since the law of the free market is not a fundamental law of nature but a false assumption of how human beings operate, there is no reason to believe it can solve the fundamental problems it has itself created.

Another proposal is to tax polluters. For instance, all producers of carbon dioxide would have to pay a carbon tax. Such taxes will allow the rich to continue polluting with impunity while the rest continue to suffer. Nor will such taxes be used to remove carbon dioxide from the atmosphere: the basic problem remains untouched.

Transfer pricing

Much of the Pearce Report is devoted to the finer points of costing – but when the basis is itself suspect, details are unlikely to yield solutions. They do, however, expose the difficulties of attempting to price the priceless.

In conscientious costing, at least all known environmental and social interconnections would be taken into account. Instead of working on the areas where "economic systems impact on . . . the global life-support systems on which we all depend", the authors are content with pricing the impact of air pollution on property values, the impact of aircraft noise on house prices, and assessing the market value of grizzly bears and big-horn sheep. The studious avoidance of the crucial question of energy use enables them to claim that net beneficial environmental results are obtainable with increasing consumption.

To cost environmental interventions, one must be able to express them mathematically. Most of the biological processes, and many of the large scale physical ones (such as the global climatic system), cannot be mathematically represented. When representations are claimed, they are based on much simplified situations which cannot give reliable results. Further, when one has lots of variables involved in lots of equations and makes lots of approximations and assumptions, one can get whatever results one wished to get in the first place. Such creative accountancy is a sad attempt at mathematizing wishful thinking.

Credit must be given to the authors for presenting many Green arguments fully and correctly. It is unfortunate that the constraints of working in an unjust society have apparently induced them to promote the contradictions of these arguments.

THE WIDER CONTEXT

Three years have passed since the Brundtland Report proclaimed that economic growth without environmental degradation was possible but till today no magic formula of how this can be accomplished has been revealed.

> The bad news is that civilisation is heading for an ecological catastrophe. The good news is that heading it off need not involve sacrificing too much economic growth. These were the broad conclusions of the conference of world experts on the implications of the world's changing climate held in Tokyo last week. However, the meeting, sponsored by the UN Environment Programme and the Japanese government, evolved no distinctively new initiatives aimed at avoiding the crisis. . . . In the long run, the conference concluded, the question of environmental destruction is intimately linked with poverty: poor people deplete natural resources, which in turn creates more poverty. The decision that wealth is an ally, not an enemy, of environmentalism, must have been immensely reassuring to the rulers of the world's richest country.[10]

So must have been shifting the blame to the victims.

The economically poor have been pauperized by the Western economic system. The poor have had their forests slashed and burned and looted by industry, for hamburgers and mahogany toilet seats. They are now being condemned for the little fuel and timber they glean for their sustenance because this diverts attention from the large-scale despoilers of the earth.

The population question

On 8 November 1989, Mrs Thatcher addressed the UN General Assembly on the subject of the global environment: "More than anything, our environment is threatened by the sheer numbers of people and the plants and animals that go with them. Put in its bluntest form: the main threat to our environment is more and more people and their activities."[11]

Rather than referring to population figures alone, it would be more revealing to use the concept of Pollution-Multiplied Population (PMP). An individual in the US consumes over 20 times the resources of the average person in the Two-Thirds World. The

US population of 250 million multiplied by 20 gives a PMP of five billion. This is equal to the present world population. The consumption of paper in the US is 137 times that of India. The paper PMP, then is 34 billion! The effect of this alone on deforestation is evident.

If one calculates the PMP, not on the basis of national averages, but in relation to the consumption rates of the poor themselves, the results are more glaring. Since neither cars nor CFCs are used by the poor, the PMP for these commodities approaches infinity for any of the industrialized nations.

The West requires a surplus of cheap labour in order to provide the cheap commodities which contribute so much to their high standard of violence to the earth. Further, a high population provides a market for their products which require increasingly scarce resources. The West is anxious to monopolize these, thereby preventing the people of the Two-Thirds World from providing for their basic needs.

There is seeming concern about the ability to grow enough food on the limited cultivable-land available. Yet millions of hectares of land are being wasted by growing tobacco, grains for the production of potable alcohol, fodder for US and European cattle, and other harmful or superfluous crops. All these could be freed for food production.

This is not to say that parents should not limit their families. They will be more than willingly do so, but only after the causes of their impoverishment are removed. Poverty compels them to have many children for support and security.

In *Sustainable Development* the author, Michael Redclift, states:

> It is sometimes argued that population pressure is the major obstacle to securing development in the South, as if limiting population could be divorced from the strategies poor people adopt in pursuing their livelihoods. Reference is made to increasing family size preventing 'human needs' being met, when large families are precisely one of the strategies open to the poor to ensure their own survival. . . . Efforts to slow down population growth will continue to be frustrated until meeting basic human needs is considered the priority of development.[12]

The equitable distribution of inequity

The Brundtland Report states: "And we believe such growth to be absolutely essential to relieve the great poverty that is deepening in much of the developing world." While this is no doubt a tempting carrot offered to encourage Two-Thirds World donkeys to follow the ways of the West, it is a cruel joke consciously played out on humankind at large by a small, selfish minority.

That the rich must get richer if the poor are to become less poor was always a suspect doctrine, at least morally; but since the extent to which the rich must enrich themselves before poverty can be abolished in this model is clearly infinite, other solutions must urgently be sought.

It is naïve to argue, as the enthusiastic proponents of business-as-usual declare, that new technology producing novel materials will eliminate both the resource and pollution problems. No technology can satisfy the large fraction of humankind who have been falsely promised luxuries in abundance if only they obediently follow the rules of the game.

Every article produced or service provided which does violence to the earth also adds to inequity and injustice. If the unjust cake which the West consumes could be equitably distributed – and there is little indication that there is a will to do so – this would merely result in the equitable distribution of responsibility for inequity. Growth, as defined by economists, will contribute only to greater distributive injustice. More people will be responsible for leaving less to the presently deprived and to future generations in a more polluted world.

We may wonder why there is such reticence about any discussion of moderating consumption. Perhaps it is because the people of the West have been pacified precisely by means of rising expectations that consumption will go on increasing. This may be what is meant by consumer sovereignty. Any suggestion that such high purposes might be interrupted is too terrible for Western leaders to contemplate. Not only would it interfere with the workings of the expanding market-economy but it would disrupt the whole rationale of Western social life. The people of the West have become addicted to eating up the earth, and with it, their own children's future. No wonder discussion of this has become taboo.

Yet the process of attaining sustainability requires the most urgent action where the problem is greatest: in the industrialized world.

We can either voluntarily reduce our consumption with all deliberate speed as a matter of abundant precaution, or wait till catastrophes happen. The small improvements that the Pearce Report suggests can, at most, delay slightly the process of retribution.

It is often said that if consumption is reduced the Western economy will collapse. However, if the survival of the human race depends on the collapse of that economy, then the sooner it breaks down the better. Where is the rationality in sustaining an economic system at the cost of our own existence?

Verbal pollution

Confusion as to our true predicament has been created by means of extensive verbal pollution. Words such as sustainability, equity and justice have been contaminated by redefinitions that belong to the rich and powerful.

The criticisms which have been deployed to block discussion of the wider implications of current economic systems are well known. These criticisms are themselves a major obstacle to the formulation of hopeful and just alternatives. It is worth examining some of them.

The search for a dynamic equibilibrium, which is what true sustainability involves, is said to be nostalgic or romantic, as though the survival of humanity were an archaic or eccentric preoccupation. The real victims of nostalgia and romanticism are those fundamentalists who believe that the laws of laissez-faire economics are the surest guarantors of human well-being. What could be more romantic than the belief that the hidden hand of free markets is the most dependable guide of human destiny? Whose hidden hand is it anyway, the vestigial trace of a God who to all other intents and purposes has withdrawn from the superintendence of our affairs, or the concealed iron fist of the rich and the powerful?

The rulers of the West cry disorder at the prospect of any fundamental change. The only permitted change is more, much more of what we have already. They insist that the people would not stand for any other form of change. If social justice were to be given precedence over wealth-creation, the whole system would collapse. Of course, the same apocalyptic predictions were made by those who wished to perpetuate other kinds of oppression – slavery, colonialism, the exploitation of the Western working-class.

More liberal-minded people will say: "Misguided idealism. We have to live in the real world". But the real world of those stern realists is an artificial construct, imposed upon us and upon the resource-base of the planet, on whose life-blood it feeds. Those who have advocated the supremacy of the bottom line must now learn that there is another bottom line: not that of profit and loss, but of what the battered world can bear.

Others assert that you cannot change human nature. But industrial society has already changed human nature, in such a way that only greed, selfishness and competition are rewarded, while restraint, altruism and co-operation have been disgraced. Evidence of the infinite plasticity of human nature is there in the very mutation that industrialism has wrought.

Many people have faith in the ability of the legions of experts, scientists, economists, politicians, to take care of all problems, while they continue their wasteful lifestyles. Those who express reservations about the likelihood of a happy outcome are accused of pessimism, as though this had become a form of moral turpitude. Yet the doom-and-gloom scenario which they reject is clearly outlined in the Brundtland and Pearce Reports. Those who accuse more rigorous analysts of pessimism must give examples of how optimism has ever averted disaster. Is it not more optimistic to believe that human beings have the capacity to acknowledge error, and exercise their free choice to correct it? After all, these are the people who are always telling us that what distinguishes the Western democracies is their freedom of choice: how is freedom of choice compatible with a human nature that cannot be modified?

Critics assert that a simple lifestyle is joyless, that those who propose this want to stop people having fun. If those who make the loudest and most joyful noises had produced a happy society, there would not be the levels of violence, fear and addiction that disfigure the rich world.

As soon as any mention is made of cutting down consumption, the monopolists of the necessities of the impoverished discover a sudden tenderness for the have-nots. When they virtuously proclaim the right of the poor to attain the living standards of the rich, they are rarely talking about redistribution. Under this cloak of convenience they are really pleading for their own uninterrupted enrichment, beyond satiation.

Yet even the apologists of injustice ("You want to bring everybody down to the same low level") – who virtuously proclaim their desire to level everyone up – are discovering as they peer into the future that the exploitation of people and planet threatens their own well-being: there is simply nowhere to hide from the consequences of their actions, no island on earth sufficiently secluded, no refuge from the contaminants and pollutants they have unleashed in the world. To leave the future of the human race in the hands of such visionaries may mean annihilation for all of us.

NOTES AND REFERENCES

1. World Commission on Environment and Development, *Our Common Future* (Delhi; Oxford University Press, 1987).
2. David Pearce, Anil Markandya and Edward B. Barbier, *Blueprint For A Green Economy* (London; Earthscan, 1989). All quotations without specific references are from this book.
3. Amartya Sen, *On Ethics and Economics* (Oxford: Blackwell, 1987).
4. Peter Spinks, "Nations fail to agree on measures to limit greenhouse effect", *New Scientist*, 18 November 1989.
5. "Science And The Greens", *New Scientist*, 30 September 1989.
6. Sarah Law, "Radio waves might safeguard the ozone layer", *New Scientist*, 16 September 1989.
7. "Ozone shield to be repaired", *Financial Express*, 3 September 1989.
8. J. Miller and Fred Pearce, "Soviet Climatologist Predicts Greenhouse 'Paradise'", *New Scientist*, 26 August 1989.
9. Fred Pearce, "A hero for the greens?", *New Scientist*, 23 September 1989.
10. "Tokyo endorses growth with greening", *New Scientist*, 23 September 1989.
11. "Thatcher stresses global effort to protect planet", *The Guardian*, 9 November 1989.
12. Michael Redclift, *Sustainable Development* (London: Methuen, 1987).

4. Technological Intervention

> Transnational corporations have a special responsibility to smooth the path of industrialisation for the nations in which they operate.[1]

The colonial exchange of wealth for poverty continued after independence. An important vehicle of transfer in this has been transnational corporations (TNCs). TNCs have inherited the genes of that pioneer in multinationals, the East India Company. They are a mutation whose characteristics even biotechnology will be unable to modify in a more benign direction.

Current patterns of development in India, like those in so many other parts of the world, seem to acknowledge the fundamental superiority of the market system. Everywhere – in Eastern Europe, in Mozambique, in Tanzania, in Vietnam – the movement is away from centrally-planned economies, with their obvious inadequacies and failures, and towards "free" markets. This is naturally seen by the West as evidence of the invincible superiority of the economic system which originates there.

The changes now occurring in the formerly socialist world are seen by the West as admission of error, and are greeted as a kind of homecoming to the classical necessities and eternal truths of Political Economy. It is perhaps understandable that the West should be so triumphalist at this time; and it should not surprise us if the general thrust of public policy in India is influenced by these same world-wide currents. Hence, public policy is towards liberalization, characterized by deeper penetration of the Indian economy by transnational capital, by growing debt (now estimated to be more than US $ 60 billion) and by dependency upon the world-system.

The very success of the West depends upon the dynamic connection of the economy with environmental destruction. In this form of development, natural treasures are merely one factor of production. The most impoverished people on earth (those made poor by development) are the infinite absorbers of the noxious by-products of the industrial way of life. It is this great contradiction that faces us in the last decade of the millennium: at the very moment when the West is celebrating its final triumph, it is running into a crisis of sustainability.

This is why India, with its tradition of self-reliance, modest use of resources and respect for the environment, might have been expected to lead the resistance to these ruinous processes. As it is, the reverse seems to be happening; and the TNCs are in the forefront of undermining our capacity to show the world the way to more hopeful alternatives.

TNCs claim to help the Two-Thirds World by transferring high technology, but in reality they send us obsolete systems that are no longer profitable or which produce items which are banned in their countries of origin. They claim to produce employment but there is net disemployment. They claim to bring in capital but they drain away far more. In this process, they are aided and abetted by their own governments and their Indian partners.

The case of Union Carbide has by now been well documented: what happened in December 1984 at Bhopal was the instantaneous destruction of thousands of lives by the leak of gas from a plant which was in the business of manufacturing poisons in the form of pesticides.

The export of dangerous and inappropriate products and unsafe manufacturing processes to the countries of the Two-Thirds Worlds is one of the major corporate crimes of the twentieth century. Thus US Congressman, George Miller, testifying in the 1980 hearings on hazardous exports: "Just as military imperialism was justified for its 'civilising' effects, so the dumping of dangerous industries and products has been rationalised because it stimulates economic development and modernisation."[2]

TNCs suddenly become eager to transfer their technology when strict pollution laws make it is no longer profitable to use that technology in their own countries of origin. The US trade magazine *Chemical Week* reported that "US chemical firms spend 44% less on pollution control at their overseas plants than at those inside the country".

It is clear that workers in the Two-Thirds World are regarded as beings of a different order from those in the USA or Europe. Management at Union Carbide clearly perceived workers as just another raw material, undifferentiated from the gas which killed them. They had gone so far as to instruct workers "to drink six or seven glasses of milk a day and go in for a high-protein diet of fish and eggs".[3] This reveals that the workers were probably being continuously poisoned, even in the normal operation of the plant. Further, the local hospital had been liberally endowed by Union Carbide, which may have led its doctors to close their eyes to the state of health of the workers they examined.

Yet, on a less dramatic scale, such accidents are occurring everywhere in the many thousands of chemical factories all over the country. In Maharashtra, which has over 40 per cent of the pesticide factories in India, the number of prosecutions has been minimal. For many years the people of Chembur in Bombay (a district which led to its being given the facetious name Gas Chembur) have been protesting against the pollution by several factories (including a former Union Carbide plant now operated by an Indian company), yet little action has been taken. All over India, factories continue to pour out wastes, which injure and harm people, and even kill them.

Pesticides are one of the most baleful commodities introduced by TNCs into the Two-Thirds World. The number of people who died or were affected in the Bhopal accident is horrifying; but this figure is negligible when compared to the multitudes being poisoned or killed by synthetic pesticides every year. The World Health Organisation figures show a million poisonings and 20,000 deaths a year.[4]

Just how many other millions of people all over the world are being hastened towards a painful and avoidable death by the slower ingestion of those poisons, which have entered the food chain and contaminated water sources, cannot be calculated.

These WHO figures are underestimates, because so many pauperized farmers have no access to doctors or medical care of any kind, and simply suffer and die without even having the doubtful satisfaction of being included in statistics.

Inas is a small farmer, about 60 kilometres north of Bombay, who grows vegetables for sale, most of which come to the city. He is educated, having completed his school studies. His children, also educated, have well-paid jobs in Bombay. Inas sprays his growing vegetables with pesticides. He is not particular about the type of pesticide he uses, and applies whatever is available in the local store. He has been told by pesticide salesmen to spray regularly every two weeks; and he does this religiously. Since he plants traditional varieties of crops that do not ripen at the same time, in order to sell them in small lots, the produce frequently gets sprayed immediately before harvest.

Actually, it is his wife who does the spraying and dusting. While spraying, she just ties a piece of old cloth around her nose, and the same cloth is used over again without frequent washing. She wears her normal clothes when spraying and does not wash them before using them again. No manufacturer's leaflets are available in the local store, though some are supplied with the pesticide sold, but Inas has never read them carefully. The print on the pesticide containers is microscopic and its dangerous message barely decipherable. When told that it is dangerous to spray without protective equipment or to spray too much or just before harvesting, he would answer "this is the way everyone uses pesticides around here". But Inas is one of the lucky ones; he has recently learnt the value of botanical pesticides and has ceased using synthetics.

Pesticides manufacturers claim that their responsibility ends with giving instructions on the precautions to be taken while using the insecticides. Most of the container labels state that the sprayers should not be washed near wells and streams and that all residues should be carefully buried and not thrown into the nearest *nalla*. But the majority of impoverished farmers wash directly into water sources which get polluted and kill or harm people, fish or other wild life. And the empty tins are sold for re-use which means that they could be used for storing somebody's food or carrying someone's water.

Consider this as an example of inadequate information. Thimet (Cyanamid India Ltd) is second highest in toxicity among all pesticides being currently used. Yet Cyanamid does not even mention what its chemical composition is (it is an organophosphate) in its leaflet "Thimet 10-G to Protect Your Cotton Crop", and there is no mention of its hazards, or the safety precautions to be taken in its use. In another leaflet, "Grow More Paddy with Cyanamid Insecticides", it is stated: "Do not use Thimet 10-G when fish are cultivated in paddy fields". Are the manufacturers ignorant of the fact that fish live in nearly all paddy fields and that field water overflows or infiltrates into rivers?

Hindustan Petroleum's Finit Household Insecticide contains Malathion, a dangerous chemical. The container states "Under the Insecticide Act 68, the tin contains space spray. It means it can be safely used in (closed) space . . . Spray in a closed room until room is filled with mist. Open the room after 15 minutes." It cautions: "Remove bird cages and pets before spraying. Avoid excess inhalation." However, nothing is mentioned about using face masks. Who would buy Finit if Hindustan Petroleum stressed on the container that you would also have to buy respirators and wear protective clothing?

Baygon Spray Household Insecticide (Bayer's) contains a carbamate that is five times more toxic than the Sevin of Union Carbide. The container states "Caution: Avoid contact and inhalation. Do not spray on foodstuff, utensils, bird cages, aquaria or humans".

In Bombay, every time the municipal personnel come around to spray, people complain of tightness in the chest, difficulty in breathing, tiredness, burning of the eyes and skin. Spiders and lizards disappear after a few days, poisoned by the insects they eat. But mosquitoes continue to bite, having developed resistance to the insecticide being sprayed.

Professor S. A. Bannerji of the Institute of Science, Bombay, has analysed hundreds of samples of vegetables, fruit, fish and meat, bought in the local markets. He writes: "Organochlorine pesticide residues are present in almost every food article that one buys in Bombay. Except for fruit and prawns, the percent of other food samples containing residues . . . was appreciably high, ranging from 40 to 70 per cent . . ."[5]

Samples of bottled milk in Maharashtra contained more than seven times the permissible limit of DDT and more than 100 times the limit for Dieldrin. "Permissible" limits do not mean that they have proved to be safe limits. DDT, Malathion and Lindane have been found in 25 per cent of the samples of food in Calcutta, with 37 per cent of the samples exceeding the WHO limit. Butter, ghee, oil, grains, pulses and even baby food have been found to have excessive contamination.

The level of DDT and BHC in the body of the average Indian is among the highest in the world. The effect of this is explained by Dr Bannerji: "Since these pesticides have been shown to affect some of the enzymes of key metabolic pathways, and since they are stored in fat deposits and are biostable, their continued intake through food could build up to dangerous levels in human tissues and ultimately affect their health and well-being." The carcinogenic effects of various chemicals in combination are incalculable; but there is little doubt that a significant carcinogenic burden accumulates over the years. The virtual epidemic of cancer in the West (in some countries one in three persons will suffer) gives rise to a great charitable activity in the raising of money to find cures. It might well be pertinent to ask whether collecting money can ever cure what the making of money, through intensified chemicalized agriculture, has caused.

The case of indigo: the "development" paradigm

For centuries indigo was obtained from a number of wild plant species growing all over India. The process of making indigo is an instance of researched technology. It required the fermentation and subsequent oxidation of chemicals in the plants, in order to convert them to the dye. Most production went for local consumption, with probably only a small surplus exported via the Arabs to Europe.

In the 18th century, access to natural dyes was crucial to the European textile industry. The demand for the dye was so great that the British saw in indigo an item for monopoly trade and a vast market ripe for development. Farmers here were compelled to change over from cultivating food crops to indigo. But monocropping of indigo was not profitable, as the Indigo Commission of 1861 recognized. Coercion was then necessary in order to force peasants to use their labour and land to grow indigo. Some were beaten to death if they did not do so, but there is no record of any aggressor being convicted of murder.[6]

But already by the 1850s British chemists were trying to find artificial substitutes for indigo. They were beaten to it by German chemists with their aniline dyes made from coal tar. In 1897, the Badische Aniline Company was mass-producing synthetic indigo.

That same year, planters in Bihar had 574,000 hectares of indigo in the ground. With this technological progress, the demand for natural indigo dropped. By 1911, the cultivated area had fallen to 86,600 hectares and many displaced indigo labourers starved to death.[7]

Biotechnological research is now being carried out on developing micro-organisms that will produce indigo in industrial reactors at lower costs than dyes produced from chemicals. Those employed in the chemical dye factories will then be dis-employed while a smaller number will find employment in the new form of production.

Most steps of technological progress result in regress elsewhere. Improvements in efficiency reduce the labour required. It is not merely that employment is transferred from one country to another, but that total global employment is always reduced for the production of a particular item. The claim that the dis-employed can be taken up by new industries is false, because of the limits to resources and limits to the pollution that the planet can bear. Further, since technology is the plaything of the rich, it is the impoverished who are further marginalized by each step of progress.

TRANSNATIONAL COMPANIES

Throughout the 1980s, the evidence of the malign effects of TNCs and their Indian collaborators and subsidiaries has grown. The Bhopal tragedy was only the most dramatic consequence of patterns of development in which the riches of India continue to be siphoned off as surely as if the country were still officially under foreign dominion. Transnationals themselves create a kind of aerial super-geography, which exists over and above sovereign nation-states, and conveys wealth from one area of the globe to another, independently of the volition, or in defiance of the legislation of mere governments.

Hoechst

A news item in 1986 declared that:

> ... prolonged court cases have made a mockery of the Drugs Price Control Order of 1979, with several companies continuing to charge arbitrarily fixed high prices for the past five years, having obtained stay orders against the prices fixed by the Government. ... The Government fixed a price of Rs 1,810 a tonne for baralgan ketone produced by Hoechst. But the company has been charging a price of Rs 24, 735 a tonne for its product, and has continued to charge the same price after obtaining the stay order. On a production of 12 tonnes over the past four years, the company has reaped an excess realisation of Rs 27 crore [1 *crore* equals 10 million] on just one item.[8]

A few days after this report appeared, Hoechst advertised in the Times of India:

> This is HOECHST in India! Hoechst began operations in India in 1956, and has since emerged as a responsible corporate structure whose four pillars are its result-oriented professionalism, its pursuit of excellence, its obsession with research, its high quality products. These, indeed, are the major objectives which Hoechst holds as its commitment to the country and its people. Hoechst is pledged to follow Government policies. ... Hoechst India Limited is already standing today at the gateway of tomorrow.[9]

After years of delay, the Supreme Court rejected their claims, declaring that "profiteering. . . in life saving drugs" is "diabolical".[10]

This is no doubt an example of what Hoechst and all the other TNCs mean when they claim to be responsible corporate structures. Since "Hoechst believes that today's record is but tomorrow's standard", we can assume that they have been following the same standards during their 30 years' stay in India. And we can estimate that their "commitment to the country and its people" has impoverished us by hundreds, if not thousands, of *crores* of rupees.

Lever

Brooke Bond India Ltd was taken over by Hindustan Lever Ltd (HLL) in the mid-1980s, thus giving the Anglo-Dutch Unilever Group, the parent company, 19 per cent of the total world market in tea. In India, HLL has 75 per cent of branded tea sales.[11] In 1987, the Tea Board claimed that the Unilever Group "is using its buying power to deliberately keep down the return to the tea growers while at the same time retaining a high profit margin through retail sales".[12]

G. K. Lieten, in his book *The Dutch Multinational Corporations in India*, shows how HLL operates. TNCs have Indians on their boards of directors, and they – and all their employees – are expected to act as "nationals" of the multinational. A 1955 report of Unilever states: "If our Indianized company is to remain a Unilever company the men and women in it will have to be Unilever people first and foremost".[13]

Lieten reveals HLL's government connections:

> Some of its directors have earlier served as high ranking directors with the Ministry of Food or as a member of the Planning Commission (particularly J. S. Raj, H. C. Bijawat and V. G. Rajadhyaskshya). Others, after leaving the company, went to occupy important government jobs. The first Indian chairman, P. L. Tandon, after retirement . . . became president of the National Council of Applied Economic Research and the chairman of the Government Task Force on Free Trade Zones. Hindustan Lever's Dr S. Varadarajan . . . in 1981 became the secretary of the department of Science and Technology and later the Director-General of the Council of Scientific and Industrial Research. Even acting managers of MNCs function on important government advisory and funding councils.

Why, one might ask, is there so much traffic in both directions between government and higher reaches of the topmost TNC in India?

HLL's consequent power over the government is shown by the fact that whereas all foreign companies had to dilute their holdings to below 40 per cent if they wished to enter certain fields, HLL had, according to Lieten, "insisted on maintaining a foreign majority in the share capital, and in 1982 ultimately got the government approval".

In order to do this, HLL had to show that a certain percentage of its production was in the priority sector and that its exports were over 60 per cent of total production. But this was not so. HLL therefore transferred foodstuff production to Lipton, its sister company. Even here there was manipulation: HLL leased only its domestic edible oil business to Lipton, keeping the export business for itself. In this way, its exports would appear as a much larger proportion of its total production.

In the 1970s, HLL had received great praise and, in addition to tax concessions, the goodwill of the government for processing *sal* (*Shorea robusta*) seed as a substitute for cocoa butter and for the production of soaps. HLL claimed: "Sal seed which has been rotting on the forest floor for centuries has now changed the health of the tribal economy".

However, the technology for extracting *sal* seeds was not developed by the company, but by the state-run Regional Research Centre at Hyderabad. The technology was transferred to the Orissa Joineries and Contractors Pvt. Ltd. which was probably the first to extract *sal*-seed fat commercially. HLL came on the scene much later. Further, the utilization of the seed for oil production was first undertaken by the industries run by the Khadi and Village Industries Commission (KVIC).

Moreover, the seeds were certainly not "rotting on the forest floor for centuries". *Sal* seed has been collected since ancient times by Adivasis who have their own traditional system of oil extraction. *Sal* oil is their only cooking medium. Before commercialization, an Adivasi family would extract about 30 kilograms of oil per season which would suffice for their cooking purposes. Nowadays, to compensate for their shortfall, they have to purchase from the market Dalda cooking oil, a product of HLL, at a price which is at least 20 times that which they get for one kilogram of unextracted *sal* seed. One can therefore hardly conclude that the *sal* chapter of HLL has helped the Adivasi economy.

There was recently a move to allow HLL and other TNCs to take over the collection of minor forest produce, particularly oilseeds, so that it can be made more efficient. If this were to happen, hundreds of thousands of Adivasis would suffer. By encouraging the Adivasis to collect more oilseeds by paying higher prices, the shortage of oil in the country might be solved, but the Adivasis would be deprived of the use-value of the produce.

HLL's dividend in 1988 was 25 per cent, with sales of Rs 1,002 *crore* and a gross profit of 94 *crore*. It is significant that in not even a single year between 1970 and 1981, has Unilever as a whole had a higher rate of profitability than HLL. In most of the years, profitability rates here were twice as high. This was despite the methods used by TNCs to show lower profits in their subsidiary companies, by such devices as the payment of royalties, transfer pricing and other equally creative manoeuvres. Unilever UK has a 51 per cent stake in the company, and this proportion of the profit goes abroad.

The tentacles of the Unilever Group spread far and wide. It is the largest plantation owner in Africa, where extensive cash-cropping is partly to blame for the extensive famines in the 1980s. Unilever has also been destroying large forest areas in the Solomon Islands. "Levers' logging in Gizo, Kolombangara and North New Georgia destroyed in a few years the accumulated capital of thousands of millions of years."[14]

In 1985, HLL was named winner of the top award by the Government of Maharashtra for the third year in succession for the company's export of *sal* oil, soya extractions, animal feeds, packet tea and other items, valued at Rs 12.98 *crore* during 1984-5. The Government has also awarded the company a merit certificate for its exports of glycerine, ossein, di-calcium phosphate and processed castor-oil, totalling Rs 21.5 *crore* during this period.[15]

The *sal* oil and glycerine can easily be turned into soap before export, but if HLL did that, its profits in India would increase further and be taxed here. Our cattle are thin and give little milk because of the lack of fodder supplements, yet HLL continues to export animal feeds. Of course, part of this comes back to us in the form of skimmed milk powder and butter oil, generously donated by the countries that feed their cows on our cheaply-exported animal feeds – the most cumbersome and expensive machine for creating the need for charity that human ingenuity could possibly devise.

Then again, phosphates are desperately required by our farmers to increase their food output, particularly of pulses, since they are essential for the rhizobia that fix nitrogen. But HLL exports di-calcium phosphate because they can make more money that way than by selling it to impoverished farmers. HLL is, in effect, exporting our sustainability.

Proctor and Gamble

A news item in 1985 said that Richardson Hindusthan (now acquired by Proctor and Gamble):

... has planned to continue its rapid growth through the development of natural products based on the ayurvedic system of medicine. To this end, the company established a full-fledged Natural Products Research Centre at Kalwe near Bombay in 1984 at a cost of Rs 1.2 crore. ... Gurcharan Das, the company's President, when asked why an international company is interested in a 2,000-year old system of medicine, stated "We have always been involved with herbal medicines. Our oldest product 'Vicks Vaporub' which is sold in 150 countries worldwide, is in fact an 'ayurvedic' product whose ingredients appear in ancient ayurvedic texts".[16]

Vicks is sold worldwide and in practically every village in India. Not only does India get nothing at all from sales "in 150 countries", but even the profits made in India leave the country. In 1984, the company declared a dividend of 35 per cent on its shares. Indian knowledge, developed over centuries by our ancestors' ingenuity and experimentation – without the use of grandiose research centres – is now being expropriated by Richardson Hindusthan to impoverish us further.

Other TNCs are also making money from our ancient remedies. The Research Director of Hoechst "pointed to the research being done by the company's scientists, to isolate natural products from indigenous plants or micro-organisms and use these separated components for the creation of novel drugs through chemical synthesis".[17]

These companies learn the secrets of our remedies and patent them here and abroad. After they are patented, no one else will be able to manufacture them here, and none of the profits from foreign sales will come back to India. Our genetic heritage is exchanged for a modern version of the colonizers' glass beads.

Bata's barter

Supporters of the Western industrial system say that if production is reduced, unemployment will increase. Yet it is that same system which by more mechanization is throwing people out of work all the time. A glaring example of this in the Two-Thirds World is the shoe industry where labour-intensive production has given way to the factory system. Like so many other basic necessities, these are no longer made locally.

There were originally thousands of cobblers making shoes to fit each individual in every village in India. Now there are Bata India Ltd and a handful of other large manufacturers which produce shoes to fit "standard" feet. In 1987, Bata produced 65 million pairs of footwear.[18] The 18,500 people employed by Bata, together with the managerial staff and shareholders obviously benefit. But their higher income is derived from the transfer of wealth from the cobblers they have deprived of a living. One of them, Lokare in Bombay, takes about two days to make a pair of shoes, less for sandals. He can make about 200 pairs a year. The production of 65 million pairs would keep about 300,000 cobblers employed for a year. To put it another way, more than 300,000 cobblers have been put out of business and many more pauperized by Bata alone. Lokare says that he has been badly affected, as he gets orders for only 75 pairs a year. If fully occupied, Lokare would earn a reasonable income. The dis-employment factor of the shoe industry is about sixteen: for every person it employs it dis-employs 16. The factory system, undoubtedly efficient in production, is even more competent in dis-employing traditional craftspeople.

The repercussions involve vast hidden costs and penalties elsewhere, which simply do not show up in the existing accounting system. When they first came on the scene, factory-made shoes were sold at prices lower than those charged by handworkers. Having allowed enough time to pass for skills to disappear, prices of shoes have shot up. Cobblers in Bombay still sell you a made-to-measure pair for less than the cost of ready-made ones. But in London, hand-made shoes are a luxury few can afford. The system not only produces unemployment, it also reduces quality and increases costs.

The manufacture of industrially-made shoes consumes lots of polluting energy, against the non-polluting manual energy cobblers use. Formerly, the skins of dead animals were tanned and treated in every village. The little pollution produced by processes that used only natural plant materials was easily handled by nature's anti-pollution systems: the biological agents that degrade natural chemicals. Modern leather-processing uses metals like chromium which are highly toxic and cannot be removed by nature. The concentrated pollution produced by tanneries is one of the main reasons for the fouling of the Ganga and other places. Traditional shoe-making is economically more efficient when all "externalized" factors are taken into consideration.

Revalorizing old skills can lead to better quality and better equality at the same time. Buying hand-crafted articles gives each individual the power to produce more employment now, while safeguarding the environment.

Multinational malpractices

The chairman of Hindustan Cocoa Products (a subsidiary of Cadbury) is actively promoting the planting of cocoa trees in India. Demand in the developed countries is increasing at 5 per cent per annum and it is claimed that India can earn much foreign exchange if more is grown and exported.[19]

Hindustan Cocoa took the initiative in propagating cocoa cultivation in 1965. One of its contributions was that as soon as the world price of cocoa went down, the company stopped buying from Indian growers and imported all its needs, thus causing heavy losses to farmers who had expected continuous sales.

Cocoa plants take three to four years to become productive and reach their peak in about ten years. In ten years time, a synthetic cocoa flavour that has already been developed, using biotechnology, will be on the market. This will displace natural cocoa, just as artificial sweeteners have already replaced cane sugar and have caused a collapse in the sugar market of Two-Thirds World countries. Indeed, this is a development likely to gather force in future years: the laboratory creation of "nature-identical" substances will, to some extent, free the rich world from its need for Two-Thirds World natural products, thereby subordinating us even more.

With monotonous regularity, Colgate Palmolive (India) Ltd. declares annual dividends on its shares of more than 50 per cent. The dividend for 1984 was 85 per cent. This high rate of profit – one of the highest in India – is partly due to the fact that much of its manufacturing is carried out by sub-contractors, where the wages paid to its workers, especially women and children, was as low as Rs 8 a day in 1985.[20] It is also in the business of displacing traditional means of cleaning teeth in India, which cost nothing.

ALTERNATIVES TO TNC PRODUCTS

Because Indian companies producing alternatives do not have the resources to advertise on the same scale as TNCs, their products are less well-known. Many of the TNCs hide behind Indian names, and fail to mention the origin of their products.

Vicco's Vajradanti and Balsara's Promise are two Indian toothpastes. Both these manufacturers have at one time been taken to court by Colgate, in an attempt to prevent them manufacturing. Balsara and Co. had to fight up to the Supreme Court to win their case. Balsara also makes toothbrushes.

However, it would be better, particularly in the rural areas, to use *babul* or *neem* twigs for brushing teeth, rather than pastes and brushes. It has been confirmed that the use of *neem* is not only good for the teeth, but also prevents oral cancer.

Indigenous soaps include Kutir (KVIC) and Chandrika Ayurvedic soap (S. V. Products, Kerala). *Shikakai* and soap nuts, still available in Bombay's markets, can be used as substitutes. Godrej and Tata also manufacture soaps, but they are themselves already vast corporations.

The East India Company was "nationalized" by its home country, not by the country which it victimized. Perhaps modern governments may find it necessary to take similar action when they discover how fragile is the allegiance of TNCs to national entities.

SOCIAL INTERVENTION

TNCs are not alone as carriers of the technological culture. Every technological advance in complexity must always be implemented, regardless of the social or the economic consequences.

The grass press

The Konkan area north of Bombay has a climate and soil that produce highly palatable and nutritious fodder grasses. These are much in demand by milk producers in Bombay and other places. Soon after the monsoon, the grass is cut and sold to the owners of baling presses; the baled grass is then transported by truck to the dairies. The grass needs to be compressed and baled in order to keep transport costs low, since unbaled grass occupies too much space for the trucks' carrying capacity.

Some years ago, Adivasis in two neighbouring villages, when asked what they needed in the area, said they wanted a grass press as there was none operating nearby. The closest was more than six kilometres away and sometimes the farmers went as far as 25 kilometres to sell their grass. Besides, the operators of the press often cheated them in a number of ways: they would offer a lower than normal rate; their grass would not be weighed, and they would be cheated by a misreading of the scales; calculations were deliberately falsified; the operators would tell them that their grass was of low quality, when in fact it was good. Further, they would have to hire a cart if they did not own one, spending half a day or more to sell just one cartload of grass. Having come from so far and spent so much time and money, they had no option but to take whatever they were offered. The result was that very little of the grass in the villages was sold. The people thought that if they had a press nearby which they could run themselves, this would raise their earnings.

The simple technology for the press, developed locally some decades ago, seemed very appropriate, so one was set up on the boundary of the two villages. At first the villagers were very happy, and people cut whatever grass they could to make as much money as possible.

But soon the troubles started. Brothers who had kept and cultivated their land in one piece began to fight, each accusing the other of cutting more than his share of grass. This led to plots being broken up. Further, there were disputes between neighbours. There were quarrels between the two villages, because the cattle from one would enter the other to graze. Too much grass was cut, and too little was left for the cattle to feed on, so they grew weak, became unable to pull ploughs or carts or to produce good calves. Many people had to buy back baled grass at nearly double their original sale price just before the rains. Worse, there was scarcely any permanent improvement in their lives, in spite of the increase in the amount of money passing through their hands. Most of it was obtained before the Divali festival and was quickly spent on city-made trinkets, a process which rapidly transformed their cash back to the pockets of the urban rich.

The net result was that though the people earned more money in the beginning, they suffered large losses later. The ill-feeling created between people and the insufficiency of the grass left for the cattle to eat far outweighed any benefits.

The Adivasi Mahamandal, a Government-controlled corporation, is supposed to promote the interests of Adivasis. It was given the monopoly to purchase grass during the drought period. After three years of drought, in 1987, the Mahamandal offered very high prices for grass. Adivasis cut and sold so much of their grass that a fodder famine was transferred to them from the drought areas. In 1989, following a good monsoon, the price was set so low that it became uneconomic to sell grass. As a result, although they suffered a loss of cash-income, their own animals were adequately fed. Economic loss may sometimes be attended by material gain.

Accompanying the export of grass is the export of soil fertility that goes with the nutrients in the grass.

It would be both more efficient and more environmentally sound if the dairy cattle were kept in the villages and the milk transported to the city. The manure would stay in the villages and far less transport would be required than that needed to convey thousands of tons of grass. Besides, consumers would benefit from lower milk prices, and farmers from a higher income, as well as being able to drink milk themselves.

However, Adivasis seem to realize that milk is an expensive food. They give priority to bulls rather than dairy cattle, since the former are a necessity for farming. Thus whatever little milk the cows produce is fed to the calves. Further, with the loss of cattle in the drought areas in the mid-1980s, it has become essential to breed more cattle.

A possible improvement would be for local people to breed bulls instead of selling grass. In addition to retaining soil fertility, the manure available will raise yields of grass as well as of their crops. Value will be added to the natural resources, there will be more employment in the villages, and improved ploughing and cartage. If more cattle are maintained, they can have *gobar* gas plants which will not only save fuelwood but also reduce women's and children's labour and make better use of cattle dung. Because of the demand for grass, much forest land has been cleared. The forest can be allowed to regenerate now that there is no market for grass.

It is easy to say that Adivasis should control themselves and retain sufficient fodder for their animals. But the cash economy is one of the methods by which "primitive" people are brought into the mainstream of modern development. And the more people depend on the cash economy, the more entangled in it they become, the more advanced they are supposed to be. As it is, we see the poor seduced into buying superfluities, even at the cost of a reduction in their consumption of basic necessities below essential levels. What more malign project could be conceived? The idea that what the people want is sacrosanct begs too many questions. It may be that people do want trinkets and trivia; but if the long-term consequences involve impoverishment and loss, this radically alters those wants. For they also want to survive; they want their children not to be malnourished, not to perish from avoidable diseases. And in this hierarchy of wants, the wanting of consumer goods would not even be a matter for discussion, were it not for the profits that the rich hope to make by promoting them, even at the cost of the very lives of the poorest.

The paddy thresher

Paddy threshing is done by picking up a bundle of stalks and beating off the grain. It appeared to be such hard work that a simple pedal-operated mechanical thresher was designed, built and tested. One person sitting at it could thresh a good quantity without much effort. The Adivasis found it satisfactory and borrowed it for use. But after some thought, it was decided not to produce any more of them.

All the big farmers in the area employ Adivasis for threshing. Introducing the machine would greatly reduce the number of those employed, particularly as an electric motor could be easily attached. The net effect, though reducing the labour of Adivasis who used it for their own threshing, would be much unemployment in the area. A few jobs would be created in making the threshers but many more would be lost in threshing.

It should be obvious that labour-saving machinery is usually unemployment-generating. Those who introduce the machine assume that it is someone else's responsibility to provide those displaced with new jobs. When the machinery is used for and operated by the owner, it can reduce the work-load and drudgery of those, especially women, who are already overburdened. It can enable small farmers to cultivate their land more intensively. Alternatively, it may give people leisure to expand other areas of activity.

Unemployment has constantly accompanied technological change, ever since the industrial revolution. The introduction of the spinning-jenny put hand-spinners out of work. The owners of the machinery became rich, but this money was not "made" so much as transferred from the hand-spinners to the factory owners, as was the surplus created by the factory workers themselves. Later, the spinners and weavers of India were de-employed, often by force, and cash was transferred from here to Britain.

Appropriate technology is usually defined as one which produces local employment, and is small and simple enough to be owned, maintained and understood by the user. It should process local renewable materials. But it can be appropriate only when it does not have unjust consequences, locally or elsewhere. Its social appropriateness is as important as its technical suitability.

NOTES AND REFERENCES

1. World Commission on Environment and Development, *Our Common Future* (Delhi: Oxford University Press, 1987).
2. Troth Wells, "Underhand but over the counter", *New Internationalist*, November 1983.
3. Radhika Ramasheshan, "Government Responsibility for Bhopal Tragedy", *Economic and Political Weekly*, 15 December 1984.
4. "Pesticide Action Network", Malaysia, 1989.
5. S. A. Banerji, "Pesticides and Pollution", Keemat, August 1984.
6. A. K. Bagchi, "Colonialism And The Nature Of 'Capitalist' Enterprise In India", *Economic and Political Weekly*, 30 July 1988. Also "The Behar Planters", *The Times of India*, 22 January 1887.
7. Cary Fowler et al., "The Factory Farm", *Development Dialogue*, 1988, no. 1–2.

8. "Thriving on Stay Orders", *Financial Express*, 25 April 1986.
9. *The Times of India*, 28 April 1986.
10. "Diabolical Profiteering", *The Times of India*, 20 April 1987.
11. "Thirsty Giant", *New Internationalist*, April 1988.
12. A. Sen, "Tea Board Flays Unilever Group", *Financial Express*, 9 June 1987.
13. G. K. Lieten, *The Dutch Multinational Corporations in India* (New Delhi: Manohar Publications, 1987).
14. World Rainforest Report No. 3, Australia, March 1985.
15. "Hindustan Lever", *The Times of India*, 10 September 1985.
16. "Richardson Hindusthan", *Economic and Political Weekly*, 5 October 1985.
17. "Hoechst Symposium", *Financial Express*, 13 October 1985.
18. "Bata to prune production", *Financial Express*, 31 July 1988.
19. "Good potential for exporting cocoa", *The Times of India*, 20 December 1989.
20. Jeremy Seabrook, *Landscapes of Poverty* (Oxford: Blackwell, 1985).

5. Farming Systems

G. F. Keatinge, Director of Agriculture, Bombay Presidency, wrote in a 1913 article:

> The old self-sufficing agriculture by which each tract, each village and each holding supplied its own needs is now largely a thing of the past. . . . The Bombay Presidency draws much of its food supply from outside, while it exports large quantities of cotton and oil-seeds. Its agriculture has become commercialised.[1]

The destruction of our traditional systems of agriculture and their replacement by systems geared to the market economy were part of British policy. While commercialization may have been progress for them, we were caught in a trap of dependency and unsustainability, which still imprisons us. Big farmers, TNCs and city dwellers thrive on expanding commercialization. Any area of development we look at turns out to have transferred resources from the impoverished to those who already have; and more often than not, has had ruinous environmental consequences too. Almost any crop will serve as an illustration.

TOMATOES

The supply of hybrid seeds, and other agricultural inputs, is a strategy whereby TNCs gain control of agricultural production. This happened in the case of the green revolution and is being repeated with the revolution in tomato production.

Growers

It was recently reported that farmers could not sell their ordinary varieties of tomatoes because hybrids, attractively large, red and juicy, had taken over the market. The price of ordinary ones had fallen so low that growers could not recover the costs of cultivation and transport. Farmers who grow and sell hybrid tomatoes are directly responsible for impoverishing those who do not grow hybrids – whether because they do not have the money to purchase the expensive seeds and other inputs, or because they realize the disadvantages of growing them.

Some growers of hybrids are lured into contracts with food processing industries which provide seeds, fertilizer and pesticides, and which promise to buy the farmer's produce. Farmers must bear all the risks of droughts and disease, while industry has already made profits on the high-priced inputs supplied. Moreover, if the demand for its products drops, the factory merely claims that the farmers' tomatoes are not up to standard and rejects them.

Seed suppliers

Farmers cannot use seeds of hybrid plants the following year since their progeny do not reproduce the hybrid's special characteristics. They must continue to buy from the big seed companies that developed the hybrid, providing a permanent captive market. There are a few such companies in India, most of them with foreign collaboration, but the government is actively promoting such business. As more farmers get hooked on hybrids, the price is raised and the farmers' profits diminish proportionately. Some hybrid tomato seeds are now sold at over Rs 20,000 per kilogram. A good part of the high seed price goes as repatriated profits of the seed producing companies.

Hybrids are developed solely for their size, colour and packing qualities, hardly ever for their nutritional content. The Asian Vegetable Research and Development Center (AVRDC), in Taiwan, is an internationally supported organization which claims to be working on "crop improvement" for the Asian population. Thousands of tomato varieties are being evaluated there. Its report for 1986 states: "The improvement program . . . is aiming to improve for 1) pink color at ripening stage, 2) year-round heat tolerance and high yield, 3) moderate firmness, 4) small but more than 25 g fruit size and 5) oval shape."[2] There is no mention of improvements in nutrition values. A reference to composition is only in connection with the proportion of solids in the tomatoes. The higher that proportion – whatever its composition – the cheaper it is for food processing TNCs like Hindustan Lever, to make tomato soup.

With nutritional values lower than those of ordinary uncommercialized varieties, consumers pay higher prices for less nutrition. Hybrid prices are so high – Rs 6 to 10 or more per kilogram in Bombay – that only the rich can afford to buy. The cheaper old varieties are no longer bought and distributed by middlemen. In many places the impoverished cannot now afford to eat tomatoes.

Tomatoes are also bred so that the whole crop ripens at the same time. This characteristic is necessary for farmers in the west whose sophisticated machines are too dumb to distinguish between ripe and unripe tomatoes. It is a calamity for farmers here who would prefer to sell a basket-load at a time in the local market. They are now compelled to sell their whole crop to exploiting middlemen.

This type of plant breeding applies to all crops developed by agribusiness. In the case of sweet potatoes: "Selections were made primarily on the basis of flesh color, appearance, uniformity, and specific gravity of the roots and secondarily on dry matter content and other recorded characters."[3]

Intensive cultivation

The intensive cultivation of hybrid vegetables requires high inputs of synthetic fertilisers. These destroy the natural soil organisms that fix nitrogen as well as others that contribute to soil fertility; the farmer then has to buy rhizobia and earthworms from commercial producers. The higher susceptibility of hybrids to pest attack demands large amounts of synthetic pesticides. It has even been reported that some farmers spray their vegetables just before sale because it gives them an attractive shine, effectively embalming them in poison. While farmers may get a little money from vegetables, they get and give cancer along with it.

New pests are being nurtured by the overuse of synthetic pesticides as well as by the loss of genetic variety. The hybrids all have the same genetic make-up and so will all be simultaneously susceptible to pest attack if a resistant pest develops – as has happened with practically every hybrid marketed till now. (This lesson is at least as old as the Irish potato famine of the 1840s.) When such losses occur, farmers desperately try to save their investment by pouring on still more pesticides, creating more resistant pests and increasing losses.

Agricultural journals regularly report the appearance of new pests of hybrid crops. One of them, a fungus, has been attacking tomatoes on a large scale. The affected fruits lose their natural lustre and colour, the skin becomes wrinkled and severely infected fruits gradually dry up.

The seed suppliers are not without their sense of macabre humour. "Moneymaker" was the name given to a tomato variety, though it "was the most susceptible to bacterial wilt". Bacterial wilt is the most devastating disease of tomatoes. Moneymaker has been discontinued in the West because it proved to be disastrously tasteless and unappetising.

Paradoxically it suits breeders when their plants develop susceptibility to pests since they can then introduce new varieties, at higher prices, claiming that these are immune. The new varieties soon themselves become susceptible.

The high inputs for all artificially bred crops require loans which bad harvests may make impossible to repay. Farmers thus get into a debt trap which replicates on a small scale that of the international world order.

SUSTAINABILITY

Other aspects of unsustainability are less obvious. When a crop is grown, a part of the soil fertility is incorporated into various parts of the plant. The roots usually remain in the soil, the leaves and stems are fed to cattle, burnt as fuel or directly returned to the soil. Part of what is consumed by cattle and humans in the locality also goes back to the soil. In subsistence farming, all nutrients are recycled locally. But with commercial farming, there is the continuous export of the soil's fertility because organic matter constantly leaves the locality. The soil nutrients incorporated into vegetables, grass, grains and timber, are exported from the field or forest. For sustainability, only that much organic matter can be exported as is replenished by the nitrogen and carbon naturally fixed in the region. And even then, the problem of supplying phosphorus and other essential nutrients remains.

The "import" of synthetic fertilizers to the field cannot compensate for the loss of exported nutrients. Synthetic farming gives a high initial income, not because of the inherent productivity of the technology but because the soil fertility is being mined. With continuous mining, yield and income decrease even with increasing synthetic fertilizer use. The soil becomes powdery and is easily eroded, either by winds or heavy rains. In Russia, no one likes to buy big watermelons because synthetic fertilizer makes them bitter.[4] As long ago as 1944, E B Balfour in *The Living Soil* pointed out that general health suffers when people consume food grown only on artificials.

If the harvest goes to towns where the sewage is recycled, the fertility is at least shifted to another locality. But if the sewage is dumped into the sea, the fertility is lost. It is worse if the harvest is exported from the country. The lost fertility, equivalent to the export of topsoil, represents a loss in the natural resource capital of the land. If exports continue on a massive scale, the productivity of the whole country must decline. The money value of commodities expresses only part of their worth; the most precious part being suppressed in conventional economic accounting-systems.

The oilseeds from crops like groundnuts or castor are sent to large cities for the extraction of oil. A common argument is that large oil mills are more efficient. However, while the actual process of extraction may cost less, there are many other inefficiencies which are not considered. The collection of large quantities to feed the mills requires middlemen who skim off large profits. The diesel fuel used in transport and the packaging required for the oil are a further gratuitous squandering of resources. The electricity used in the mills pollutes at the generating plant site. And the oil mills themselves pollute. More oil may be extracted from the oilseeds by modern processing methods than is possible by village *ghanis*. But what remains in the village provides extra nutrients for cattle or manure.

The high technology used in the mills requires less labour than the low technology of village *ghanis*. The transport of oilseeds to city mills also transfers jobs from rural to urban areas. Some of the packaged oil is sent back by truck or train to the villages which grow the oilseeds. The villagers pay high prices for transport, packaging and advertising, the high salaries of city workers and the profits of the mill owners. Villagers export their earnings and import poverty. Taking these costs into account reveals the efficiency of large industry to be a delusion. Some oilcakes, of groundnut, for instance, are exported to Europe in order to earn foreign exchange. Groundnut oilcake exports in 1987–88 were about 300,000 tons. Once in Europe the oilcakes are fed to cows and the milk is dried and sent back to us as "aid". The high milk productivity in Europe is due to such manipulations. The Netherlands requires animal feed from agricultural land of more than four times its total land area.[5] On the other hand, if the price of the oilcakes was not determined by the world market, but was within the reach of local farmers, they could feed the cakes to their own cows. Then there would be no need to import milk. The success of the cities is inseparably linked to the degradation and decline of rural areas. The growth of slums in cities of the Two-Thirds World provoke the same kind of horror that the expansion of Manchester or London caused in the early industrial era in Britain. Yet it has to be said that the poor of the cities, squatting, often illegally, in self-build slums, nevertheless perform miracles of survival, with resources that to the rich would seem insufficient. It is thanks to the ingenuity and creativity of the people, especially women – who are economists, architects, designers and planners in their daily lives. The survival of human beings in some of the most inimical conditions imaginable offers the greatest hope that if there were to be social justice, a dignified life for all would be possible. Of course, this would involve a redistribution of existing resources to both the rural and urban poor; but the prospect of a decent sufficiency for all can be realized, if we ally the human resources of the poor with a fairer application of those material resources with which we are currently so prodigal.

Any interference in a system which has proved its sustainability by a long tradition of success can cause harmful ripple effects. The continuing encroachment of the Western development system is pushing to the margins of subsistence those very people whose sustainable practice could offer us such wise and hopeful instruction, if only we paused to watch and learn from them, rather than compelling them into the destructive mould of development that has brought us to the edge of disaster.

International rice research institute

Large collections of paddy germplasm – representing enormous agricultural wealth – have been assembled at various centres in India and in other Asian countries. One of these collections, comprising 17,000 varieties, was assembled by Dr R. H. Richharia, in the Madhya Pradesh Rice Research Institute. In 1960, the Ford and Rockefeller Foundations set up the International Rice Research Institute (IRRI) in the Philippines.[6] IRRI aimed to develop new varieties of rice which would give higher yields. IRRI needed the Indian germplasm collections for its own research and to avoid competition. The Madhya Pradesh government was offered Rs 4.5 *crore* by the World Bank, on condition that it close down the MP Rice Research Institute and pass the germplasm on to IRRI. Richharia states: "Attempts are being made to see that this indigenous germplasm vanishes, with the least possible delay. This is being done by collecting the seeds of indigenous rice varieties from the growers in exchange for seeds of dwarf and semi-dwarf HYV's".[7]

The dwarf varieties, introduced by IRRI, absorb more fertilizer than traditional varieties and hence yield more, without toppling over because of their short stems. But their short stature makes them less able to compete against weeds, increasing the need for herbicides. They also produce less straw for cattle fodder or thatch. Some high yield varieties (HYVs) smell badly when cooked. Some traditional varieties are scented, others more nutritious and tasty, cook faster and fetch a higher market price. They are more able to withstand disease and drought than the new varieties.

HYVs can be propagated by seed collected by farmers themselves, which limits commercialization. But susceptibility to pests readily solves this problem. The Brown Plant Hopper (BPH) became a destructive pest of paddy only after the introduction of the new varieties. IRRI scientists noticed in 1969, that the varieties IR8, IR17 and IR20, introduced years earlier, were susceptible to BPH attack. Instead of questioning the basis on which the HYVs were being promoted, in late 1973 IRRI released IR26 and other varieties with a single gene for resistance to BPH. However, by 1975, a new BPH strain, labeled biotype 2, broke out, attacking all the new varieties on a massive scale (250,000 hectares in Indonesia alone). Later that year, IRRI released IR32, which carried a gene for resistance to biotype 2. Three years later, in 1978, a third BPH strain, biotype 3, attacked IR32. By 1978, planthopper populations had also become resistant to a number of insecticides.

It was further found that some insecticides actually increase BPH numbers. Methyl parathion, fenitrothion, decamethrin, diazinon, and fenthion cause female BPHs to lay more eggs. Scientists found that plants sprayed with deltamethrin subsequently become more attractive to whitebacked planthoppers, and numbers build up again more rapidly than on unsprayed plants. Not using pesticides results in less damage to the crops by planthoppers.

With their high requirements of fertilizer, pesticide, herbicide and water, these varieties should more appropriately be called High Input Varieties (HIVs).

Farmers in Tamil Nadu who changed over completely to HIVs have now lost faith in them. They are desperately trying to find the traditional varieties which they grew earlier. But these have also been lost or their genes preserved only by the breeders. They are now trapped by HIVs and their breeders, who are in a position to sell to the farmers what they formerly obtained freely.

Scientists at IRRI, in its first decade, thought that the varieties they developed would be suitable for use all over the world. But they soon discovered that local climatic, soil and other variations, and even taste preferences, made them unsuitable for wide application. IRRI now serves only as a collection and distribution agent for rice germplasm from national and other breeding stations. This onerous chore will ensure that all germplasm reaches commercial seed producers. No doubt, it will take another couple of decades before the slow learners in the national laboratories realize that the few varieties they can develop are still unsuitable for the wide range of ecological niches. Research will then go back to farmers everywhere, from whom it should not have been expropriated in the first place.

Return to sustainability

The changing global climate could drastically affect the output of paddy, our main source of food. HIVs are particularly vulnerable because they need water throughout their growth period. They may not adapt to changes in climate, and farmers will have to wait years until new varieties are developed in research institutions.

Local varieties on the other hand can be selected for adaptation by any farmer. It is a common custom in many places for farmers to harvest the best paddy heads in their fields using a small knife rather than a sickle. In this way they choose the seed for the next year and the crop continuously improves.

Richharia's studies showed that Adivasis had elaborate methods of selecting their seed for the next crop. They also mix varieties which produces hybrid seed, with their expected vigour, for the next crop. Hybrid rice has only recently been developed by agricultural institutes. Richharia has developed a method of clonal propagation by which a single grain of paddy can be multiplied to produce thousands of plants within a single season.[8]

The paddy ecosystem

In the Konkan region around Bombay, the duration of the monsoon is very short. Farmers use a system of transplantation that overcomes this natural constraint. They sow paddy in small seed beds which can be well manured and tended.

"Waste" lands and forests provide manures. Months before the paddy season, farmers collect cowdung, and vegetable matter from a number of plants, not all legumes. In addition to herbs and shrubs, leaves of deciduous trees that have been shed naturally, are collected. Only a few side branches of big trees are lopped. Some of the plants used are: *inodi, saag, adulsa, neem, sher, rui*. The last is particularly used for saline lands. Such trees often grow or can be planted on bunds and hedges.

The materials collected are spread on the surface of the seedling field, covered with a thin layer of soil and burned slowly, just before the monsoon starts, in a process called *rab*. Although, in burning, all the nitrogen in the organic matter is lost, *rab* is a quick method of providing other nutrients. In particular, it gives potash which is usually the limiting nutrient. Paddy varieties which would normally take a long time to mature thus get a vigorous start, compensating for the short monsoon period. Normal decay of organic matter incorporated into the soil would not be fast enough to help the seedlings. Composting is difficult because much water is required. The process also helps to kill weeds and their seeds, and harmful organisms, as well as conditioning the soil itself.

With the first rains, the seeds are sown. The farmer then prepares the main fields, whose soil is too hard to be ploughed when dry. This ploughing incorporates into the soil all the weeds that grew in the dry season. The growth of leguminous weeds between harvests is encouraged by leaving areas adjacent to the fields to run wild. This provides seeds without any labour. *Mundi*, a non-leguminous, dry season weed in paddy fields is traditionally considered good for manure. This is probably because its roots go deep and bring up nutrients from lower soil layers. *Vikhari takla*, a self-seeding legume, grows up after the harvest in paddy fields and stays green as a cover crop during the dry season. Another ploughing is carried out just before transplanting, which incorporates weeds that come up with the rains. Constant submergence of rice fields at certain stages helps to control the weeds.

In Tamil Nadu, where farmers are troubled with grasses that look like rice, a purple-leaved variety of rice is sown every five years, although its yield is much lower than normal varieties. The seedlings can be easily distinguished and the weeds eradicated.

HIVs require large quantities of nitrogen fertilizer. This is more so because only up to 40 per cent of what is applied is taken up by the crop. The rest is leached out to pollute river and ground waters. Further, nitrogen fertilizer enhances planthopper populations, increases the severity and incidence of paddy blast, of brown spot, and of bacterial blight in paddy.

Green manuring with the leaves of *dhaincha* and other *Sesbania* species increases the efficiency of use of nitrogenous fertilizers. *Neem* cake, when used for soil amendment or added to urea or ammonia-containing fertilizer, lowers nitrogen losses by inhibiting denitrification while enriching the soil with organic matter.

Traditional varieties, and even some HIVs, can get sufficient nutrients with organic manures, crop rotation and the natural nitrogen-fixers in paddy fields. Whatever cowdung is available is also used and cattle are allowed to graze on the herbs and grasses that come up with the first rains, converting them into nutrients in their dung and urine.

Green-manure crops of fast-growing legumes like *tag* and *dhedhar* are sometimes grown on the main paddy fields. These grow up to 20 centimetres within the three weeks between the first rains and transplanting time, before which they are ploughed in. When thickly sown, *dhedhar* forms a covering mat which smothers weeds.

No cash outlay is required for green manuring with *dhedhar* and *tag*, and the labour involved is mainly for collecting the seeds and for a little extra ploughing. The green matter in paddy fields helps in other ways. In the decay process, the carbon dioxide produced is used by the layer of blue-green algae on the soil surface. The algae, in turn, release oxygen which is taken up by the roots of the paddy plants. The manure also improves the texture of the soil and helps it retain moisture during periods of drought.

Some species of blue-green algae, which grow naturally in paddy fields, fix atmospheric nitrogen. *Azolla*, a beautiful, tiny fern that floats on the surface of stagnant water, grows in symbiosis with *anabaena*, a nitrogen-fixing alga. *Azolla* grows very fast, producing up to one ton of green biomass per hectare per day, containing up to three kilograms of fixed nitrogen. The use of *azolla* can provide yields as high as those obtained from chemical fertilizers.

The use of chemical fertilizers kills these plants or reduces their nitrogen-fixing efficiency. It may, therefore, be necessary to reintroduce them in certain cases. Cultures of *azolla* and blue-green algae are available for use as starters.

Azolla is usually incorporated into the soil by beating it down with twigs or by draining the field. An old method used *adulsa* to do this with minimal labour. Leaves and twigs of *adulsa* were thrown on the surface of paddy fields since chemicals in *adulsa* kill lower aquatic plants. Duckweed (*alemna* species which is also rich in nitrogen and phosphorus) would also be incorporated into the soil by *adulsa*.

In some places, paddy and *tag* seeds are sown together. When the plants have grown, the field is lightly ploughed and a kind of harrow is passed over it. The paddy plants mostly recover, but the tender *tag* is buried underground and dies. The few *tag* plants surviving are removed at the time of weeding and buried in the soil.

Mahua and other leguminous trees, which can stand waterlogging, are allowed to grow in fields so that bird and fruit bat droppings provide fertilizer. These are more advantageous than green-manure crops since they leave the fields free for other crops, are productive throughout the year, bring up nutrients from deep underground, and provide fuel, fodder and other items. Their elimination has been partly due to the use of tractors which require the removal of such obstructions.

Flooded rice fields store an enormous quantity of water, releasing it slowly into *nallas* or allowing it to seep underground. They play a large part in regulating water flow after heavy rains. But some foreign farming experts have recommended that paddy be replaced by more productive crops in the tropics.

Paddy yields have been sustained by preserving the complex ecosystem of the flooded fields and their surroundings. A paddy field has algae, *azolla*, insects, fish, frogs, crabs, birds, and other creatures, weeds and trees, all in webs of interdependence. As long as this micro-ecosystem is not interfered with, the natural fertilizing and insect control processes enable a paddy field to yield steadily for thousands of years.

Fish eat small aquatic plants and insects, including mosquitoes and their larvae and other pests. Their droppings provide instant fertilizer. As they grow, some species swim away to *nallas* where they are caught and eaten. Further, when the water dries up, those that are not eaten or washed away into the *nallas*, die and provide additional fertilizer. Chemical pesticides kill fish and other important creatures.

There are about 80 species of insect parasites and predators in rice fields. A large variety of birds also live off paddy-field insects. These are sufficient to take care of plant and leaf hoppers and most other pests. The important BPH predators are spiders and water striders. A single wolf spider in a rice field can eliminate at least 20 BPHs in a day. Leaf-folders are controlled by several insects, including predatory beetles. A tiny wasp parasite lays eggs inside leaf-folder larvae. Another wasp parasitizes the eggs of the black bug.

Farmers have many methods of reducing pest damage. The smoke from *mahua* oilcake is used on paddy blight. Some pests are eliminated by flooding fields for a day or so. To control thrips, the rice nursery is irrigated so as to submerge plants for some time and then the water is drained to wash away the insects.

Other pests are destroyed by putting the sap of particular plants, such as *kachoo* and *bihlangani*, in the water inlets to the fields. Some are eliminated by using buffalo dung and urine diluted in water, or ash and water. The latex of *Euphorbia* species is used in a similar manner. Bamboo leaves are buried in inlets if paddy plants turn yellow and this restores them to health. Religious custom requires farmers to put up lamps at certain times of the year, which kills nocturnal pests.

Crabs are a nuisance since they make holes in bunds through which water drains. Farmers pour a mixture of cowdung and water into the crab hole, forcing the crab to emerge. *Karanj* leaves, cut into small pieces, are sometimes added to the cowdung.

In some areas, ducks are allowed inside the fields after the harvest of the short-term rice crops, to eat snails and insects in the stubble. When mealy bugs attack, the spot is burnt after the harvest to prevent recurrence of the pest. *Rui* leaves are used as green manure to control this pest.

Trees in or near fields, whether leguminous or not, provide perches for birds, shelter and nesting sites, and alternate food in other seasons. Bamboos and other plants are staked in the fields for the same reason.

Ambadi seeds are sown intermixed with rice in upland dry paddy fields for controlling termite attack. *Lal ambadi* is considered better than *ambadi*. In Tamil Nadu, farmers plant, in every tenth row, a variety of rice which is highly susceptible to stem-borers. The insects feed only on these rows and leave the rest untouched.

Neem can be used for the control of several major pests of paddy. Stem-borers will starve rather than eat plants treated with neem extracts. When paddy is sprayed with *neem* oil, the number of BPHs is reduced and the pest fails to transmit the grassy and ragged stunt-viral diseases. It also prevents the transmission of the rice *tungro* virus by green leaf-hoppers. Spraying with *neem* deforms the body appendages of the rice ear-cutting caterpillar. The rice leaffolder larvae develop abnormalities within 24 hours of treatment with *neem*. The number of eggs laid by the rice leaf-folder decreases, their hatchability declines, while the rate of parasitization increases.

Neem does not kill the natural enemies of plant hoppers and leaf hoppers. In fact, paddy fields treated with neem showed a higher population of natural enemies than untreated fields. The use of *neem* does not make the grain bitter.

While most paddy is rainfed, irrigation from numerous tanks which conserve water permit irrigation in case of monsoon failure. However, these small, local, systems are dying out because of governmental bias towards large scale systems and the replacement of people's control by central directives.

In spite of the short duration of the monsoon, ingenious methods of obtaining multiple crops without irrigation have been researched and are practised.

In Tamil Nadu, mixtures of three-month and six-month-duration paddy varieties are sown together. When the short-duration variety is ready for harvesting, both are cut at ground level. A special plough is then used to split the tillers of the six-month variety, which grow rapidly and provide a second crop. In another system, *math* beans and *bajra* are sown together. The *math* keeps down weeds while *bajra* is growing. After the *bajra* harvest, the *math* is left standing. Dry-land rice is then broadcast over the *math* after which the *math* plants are uprooted by hand and dropped on the soil as a mulch.

Richharia has suggested that it may be possible to grow mixtures of non-lodging, stiff-strawed types of rice with lodging high-yielding types to prevent the latter from lodging. If a suitable combination is worked out, it may even be possible to prevent lodging as well as to produce hybrid seeds by chance cross-pollination to exploit hybrid vigour.

On field bunds, *gowar, bhindi, val, tur, chaoli, nagli, vari,* or other cereals and vegetables are grown. These protect the bunds from being eroded in heavy rains. *Chaoli* is preferred because it forms an effective ground cover and also because it is ready for harvest before the main crop.

In some places, the seeds of a variety of *jowar* are mixed with rice. The *jowar* plants when about one metre high are uprooted for fodder.

After the paddy harvest, leguminous crops are grown in fields whose soil retains sufficient moisture in the dry season. One traditional system uses *tag,* in one year and either mung or *udid* in the second, but not *tag*. Others grow *val,* mung or *harbara.*

The use of these simple techniques can raise paddy yields considerably, at little cost and risk to farmers. In additon, non-renewable resources are replaced by renewable ones, and there is a considerable improvement in the soil.

While HIVs may give a higher yield for a short time, they are not sustainable. It may be said that the use of traditional varieties will not provide sufficient food for growing populations. But the use of traditional varieties will allow these yields to be obtained practically indefinitely and if more needs to be grown, it can be done by replacing such crops as tobacco with food crops.

There are, of course, a multitude of local practices throughout India and elsewhere – many already documented – some still doubtless unknown except to those who use them.

Many of the traditional practices deliberately sacrifice immediate gains for the sake of long-term sustainability. Such discipline and foresight is incomprehensible to those who practice western agribusiness since their depth of focus is limited to the next balance sheet.

THE PROBLEM OF PESTS

Until imported agricultural practices are changed to more wholesome ones, the continued use of pesticides will unfortunately still be necessary.

There are valid alternatives to synthetic pesticides, such as the use of plant-derived substances, intercropping, crop rotation, the use of traditional rather than High Input Varieties and other biological pest-control measures.

More than 300 plant species have been traditionally used for pest control in India and there is a large store of knowledge on the subject, knowledge verified by continuous practice. Many other plants may have insecticidal or insect-repellent properties. Each plant species is normally attacked by only a few insects and diseases; otherwise the species would not survive. Every plant species must, therefore, possess the ability to repel or harm those which do not attack it. This ability could be a physical feature – such as sticky sap or deterrent hairs – or chemicals which repel, kill or harm the pest. It should not be surprising then that *neem* for instance, is effective against so many insects. Since most of these plants still grow wild, they are available free to the farmer.

Agribusinesses have recently patented processes of coating seeds with pesticides and fertilizers. Such a process was used here centuries ago. Crop seeds were coated with cowdung just before sowing. This acted as a pesticide particularly effective against fungi which attack seedlings. It discouraged birds from eating the seeds after sowing, and it provided nutrients close to the emerging roots.

Neem for pest control

One of the most important plants for pest control could be *neem*. *Neem* leaves and seeds have been used as insecticides for stored grain in India for centuries. Indian scientists began research on *neem* as a botanical pesticide for field use in the late 1920s. The antifeedant properties of *neem* were discovered in the Indian Agricultural Research Institute in 1960. Since then, *neem* has been intensively researched in the West, and is being found effective against an increasing number of insects.

Neem is known to control a wide variety of pests, about 200 species of insects, mites and nematodes. It is effective in many cases where synthetics do not work. *Neem* controls pests of crops and stored produce, and some household pests. It is effective against about ten species of beetles, ten species of flies, 25 species of butterflies and their moths, nine species of grasshoppers and locusts, as well as several species of aphids, fungi, leaf hoppers, leaf miners, mites, nematodes and termites.

Neem is cheap and can be "manufactured" by every farmer. *Neem* leaves, fresh fruit, kernels, seed cake, or products such as soap, may be used. Moreover, there is no need to purify its chemicals. Crude extracts are highly effective, often more so than refined extracts, because some of the active chemicals may be removed by the refining processes.

Synthetic pesticides have a single active ingredient which affects only one particular aspect of an insect's organism: the reproductive, respiratory or nervous system. Thus, a single mutation of the insect can produce resistance to the chemical. *Neem* and other botanicals have many active ingredients, each of which may affect several of the systems simultaneously. It is extremely difficult for insects to develop resistance to botanicals.

Neem does not pollute; it is harmless to farmers and to those who consume the crop on which it is used, as well as to birds, fish and mammals. When *neem* is sprayed on plants, it loses much of its activity within a few days because sunlight degrades its active chemicals. However, some of the purified compounds of *neem* are very stable under dry and neutral pH conditions.

Neem-seed cake can be used simultaneously as a good organic fertilizer and pesticide.

How neem works

Neem contains about 45 biologically active compounds, of which around 30 affect pests. The chemicals are taken up internally by the plants on which *neem* is sprayed and they work from within. Insects that bite or chew the plant, or suck its juices, are either inhibited from feeding or are affected by the chemicals ingested. *Neem* has been called "the world's most fantastically effective natural insect antifeedant known today". The desert locust will starve rather than eat plants treated with *neem* extracts. Among the pests which are affected by the ingested chemicals are caterpillars that eat the leaves, plant-hoppers that suck juices, leaf-folding caterpillars that cut the leaves and ear-cutting caterpillars.

The active ingredients of *neem* can also penetrate the cuticle of larvae of butterflies and moths, of beetles, and of nymphs of bugs, grasshoppers, and certain other insects. Direct spraying is therefore also effective.

Some of *neem's* chemicals disrupt the stages of growth and insects may remain in the larval stage. Some chemicals cause the insects to lay fewer eggs or become sterile. Others kill them outright.

The most important chemical appears to be azadirachtin, which is absorbed by plants and has systemic action. Azadirachtin is effective even at as low a concentration as 0.1 part per million and so *neem* does not require particular care in use. Two other active chemicals that have been identified in neem are meliantriol and salannian. Several others are being tested for their effects.

Azadirachtin and possibly other constituents present in *neem* cake, are taken up by the roots of the plant to which it is applied and spread from there to all other parts.

There is no need to wait until it is known exactly how *neem* works. It can be used where it is known to control pests, and tested on other pests where it is not definitely known to be ineffective. There may be chemicals of whose effects scientists are still ignorant.

THE USE OF NEEM

Field use

To use *neem* leaves, about five kilograms of fresh leaves are put in 30 litres of water and boiled for half an hour. This is left to cool overnight and sprayed with a hand sprayer. This is sufficient for two sprayings on paddy on about 500 square metres of land. No further technology is required.

This preparation has been used successfully against the paddy stem borer. Heavy infestation was controlled within three days of spraying. It was also used successfully on a hairy caterpillar attacking *til*.

A significant proportion of *harbara* and *tur* crops are lost annually through pests. *Neem* can prevent most of the damage. Cotton crops are being devastated by boll worms and white fly. Several farmers committed suicide recently because the failure of the synthetic pesticides to control the pests led them to increasing dependency and deepening debt. Yet *neem* could have controlled these pests without difficulty.

Stored produce

Traditionally, layers of *neem* leaves, about five centimetres thick, are spread over stored grain and a paste of leaves or crushed fruits is applied to the inside walls of the earthen or cane storage containers. Farmers burn *neem* leaves for fumigation of stored paddy and pulses.

Neem oil acts as a protective coating on grains, pulses, and beans, which deters insects from laying eggs. Only a small amount of oil

(2–4 millilitres per kilogram of grain) is needed, but the produce and oil have to be mixed thoroughly.

Dried leaves or seed powder mixed with cereals gives protection for many months against the rice weevil, lesser grain borer, the red flour beetle, the angoumois grain moth, the khapra beetle and other pests. They also protect potatoes in storage. Neem cake can be used as a pesticide for stored grain. Pulses treated with *neem* do not show any effect either in taste or smell after washing and cooking.

Household pests

Seeds or seed oil can be burnt to repel flies and mosquitoes. *Neem* soap for bathing discourages mosquitoes. Extracts of *neem* strongly inhibit the development of larvae of some mosquito species; enrichment with azadirachtin increases the effect. In high concentrations *neem* is toxic to fish that eat mosquito larvae and tadpoles.

Neem delays the moulting of the common cockroach nymphs, which kills them. The leaf juice or seed oil, applied to hair, kills lice. The leaves, dried in the shade, are commonly placed in books, paper, and clothes to protect them from moths. *Neem* timber used for windows, doors and furniture resists insect attack.

Limitations of neem

The active ingredients break down in sunlight within a few days. Mixture with oils such as *erandi* (castor) and *vekhand* (calamus) give slight protection from degradation. Spraying crops in the evening will give *neem* more time to work.

Neem does not act against every pest. It can harm useful insects like bees. Spraying in the evening will reduce the harm. *Neem* leaves are toxic for rabbits and guineapigs. Ethanol extracts are toxic to snails. Spiders, ants and birds are not affected.

Too high a concentration of *neem* oil can cause plants to be malformed. Tobacco seedlings 20 days old were adversely affected in this way.

The commercialization of neem

Scientists who support the Western pesticidal industry are working frantically to identify, isolate and synthesize the active chemicals in *neem*. However, since the molecular structures of azadirachtin and the other active chemicals are very complicated, they may not succeed. Synthesizing the chemicals will allow only one or two of the many active chemicals to be applied and will not be as effective as natural extractions. However, some extractions are already on the market.

CONCLUSION

There is considerable interest in Integrated Pest Management. However, IPM appears to be an attempt by the scientific and industrial establishment to co-opt the self-reliance movement. It requires experts to observe the incidence of pests, who then continue to recommend the use of synthetics, the removal of "weeds" and so on.

It is questionable whether "pests" exist at all. It may be that our ways of cultivating crops turns benevolent species into harmful ones. Pests, then, are mere symptoms of a much deeper malaise: agricultural practices which work against rather than with nature. Plants survive under natural conditions without pesticides. Agricultural practices which tend to imitate natural conditions (organic manuring, intercropping, surroundings which provide a habitat for a diversity of creatures will probably make pesticides unnecessary.

This appears to be confirmed by just three years of organic farming in fields in the Konkan region, whose surroundings were not cleared, but allowed to run wild with shrubs, trees and even "weeds". Today, it is difficult to find a single blade of paddy which has even been nibbled at, though there are plenty of insects about.

Household pests can also be reduced drastically by letting the house "run wild". Lizards (geckoes) and spiders control flies and cockroaches effectively. Some people use snakes to control rats. Trees and shrubs around the home attract flycatchers and bats which keep down flies and mosquitoes. Planting *tulsi*, which is grown in nearly every home, repels mosquitoes.

NOTES AND REFERENCES

1. G. F. Keatinge, "Agricultural Progress in Western India", *The Poona Agricultural College Magazine*, July 1913.
2. "AVRDC Progress Report Summaries 1986", Asian Vegetable Research and Development Center, Taiwan.
3. "1985 Progress Report", Asian Vegetable Research and Development Center, Taiwan, 1987.
4. Boris Komarov, "The Destruction of Nature In The Soviet Union" (London; Pluto, 1978).
5. Anon, "Third Word Agronomists Interview a Dutch (Alter)native Farmer", ILEIA Newsletter, October 1988.
6. Michael Hansen, "Escape from the Pesticide Treadmill", Consumers Union, New York, 1987.
7. R. H. Richharia, "An Action Plan For Increasing Rice Production In India", PPST Bulletin, Madras, Apr. 1987.
8. R. H. Richharia, "Rice In Abundance For All Times Through Rice Clones: A Possible One-Grain Rice Revolution", Bhopal, 1987.

6. Interconnections of Violence

Most people connect violence solely with physical action against other human beings, but ancient Indian sages perceived it in a much wider sense.

They considered all life sacred, and in their concern for self-perfection, the killing of any living being, human or non-human, was sinful. Further, causing harm to other creatures was also thought wrong and had to be minimized, because harm itself was considered a form of partial death. Harm was defined widely to include not only physical injury, but also all forms of pain, including depriving persons of their livelihood or intimidating them. Violence could be committed personally, it could be instigated or aided, or it could be condoned by observing it without protest.[1] However, it is not possible to survive in this world without at least some violence, for we depend on other living beings for our food. Avoiding all killing results in our own death.

Our sages were deeply concerned that humans must necessarily be involved in violence and death and that absolute innocence was unattainable. They understood the concept *ahimsa* to mean the minimum or least possible violence. While causing some harm is inevitable, we do not have a licence to kill other creatures ruthlessly, to act on the basis of the "survival of the fittest", which in effect means survival of the most violent. Rather, we should have greater respect for those beings whose lives must be sacrificed in order that we may survive. The survival of the fittest means that the rest perish. Social justice is incompatible with this theory.

Gandhi extended the traditional concept of *ahimsa* further. He said that violence could also be committed by participating in or benefiting from a harmful practice. *Ahimsa* demanded compassion and love; it was not merely a negative virtue of avoiding injury to others, but a positive one of stopping harm being done to them and helping those who have been hurt. Identifying oneself with all other living beings helps immensely in putting this into practice.

The difficulty in living up to these ideals was recognized by Gandhi; but this difficulty is no reason either to run away from the world, or to give up all hope of changing it.

Modern society is extremely violent to human as well as non-human life. There is the obvious violence that one sees daily: police brutality, caste, class, communal and minority conflicts, rape and dowry deaths, abuse of children. However, much more widespread and subtle is the violence embedded and inherent in the objects and services we ordinarily use. Because most of these originate far away from us in space and time, we remain in ignorance of their content of violence.

The success of the Western economic system is claimed by its defenders to be the least coercive means of creating wealth and allocating resources. Such extravagant claims could never have been made to credulous Western peoples had it not been for the extreme brutality with which colonial empires were pressed into its service in the early industrial period. Patterns of subordination set up at that time, have been perpetuated in the ostensibly post-colonial era.

By its appeal to universal laws, whether economic or natural, this truly violent system has sought to exculpate itself from the barbarities it has inflicted upon the earth and its peoples. These barbarities were said to be the consequences of impersonal rules of supply and demand, market forces, the necessities of profit and loss. Those rich individuals and social groups who benefited from the impoverishment of others have sought to absolve themselves from responsibility for the fate both of the poor of the earth, and of the planet itself.

SOURCES OF VIOLENCE

Resource use

When jungles are destroyed billions of creatures – from elephants to tiny insects – are killed or driven away to find sustenance in other already overcrowded and degraded areas. Countless plants from huge teak trees to tiny herbs, fungi and bacteria, are being annihilated. Many species of plants and animals are wiped out completely, while still more are seriously endangered.

Harm is also done to the people – like the Adivasis – living in the forests who depend on them for their basic necessities. Besides being directly impoverished, their ancestral and spiritual links with the land which they have occupied for centuries are broken. The past and continuing history of colonialism is a story of cultural contamination and of whole peoples dying of grief at the destruction of their civilization.

All articles made from forest produce carry with them the harm inherent in forest destruction. Much wood goes for house construction and furniture. When there were plenty of trees around, sufficient firewood was obtained from dead branches and loppings. But now that so much forest has been destroyed for other reasons, the use of firewood is causing further devastation. Yet firewood is a basic necessity: it is other forms of consumption that must be curbed.

Wood is used for the production of plywood and hardboards, for paper and for packing apples in Kashmir to be sent to other distant places for consumption. Some minor jungle products have also resulted in particular species of plants being wiped out over large areas. For instance, practically all *kahandol* trees in parts of Maharashtra have been killed because they were over-tapped for their gum, which is exported for confectionery. It is said that some companies, to expand the use of their dental products, are encouraging the felling of *neem* trees so that fewer twigs will be available for brushing teeth.

Dams and other big projects

Jains have long believed that ploughing the soil is harmful since numerous living creatures are killed or disturbed in the process. How much more harmful, then, are such mega-projects as dams and other huge construction works which devastate the earth and the creatures in and on it, on a very much larger scale.

Over one million people are being displaced from just three such river projects: the Tehri, the Narmada Valley and the Bodhghat. Rehabilitation does not prevent the harm being done, it only mitigates it slightly. The banished are usually given inferior land — if they are given anything at all — and then are quickly forgotten.

Jungles are submerged in the reservoir areas while other wooded areas are destroyed for access roads and construction works. These projects also require vast amounts of funds which deprive other sectors, such as education and health.

The supposed benefits of these projects are irrigation for food and cash crops, power generation for industry and agriculture, water for domestic and industrial use and increased employment. Much of this goes for non-essential purposes. Because so much water is already used for cash crops and is wasted by industry instead of being recycled, more dams are now required for food. Those who benefit from the dams for non-essential purposes become guilty of violence. But the benefits do not go only to the farmers whose land is irrigated, to the industrialists who set up factories and the people who get employment. They extend also to those who use the farm and industrial produce as well as the power generated. With each benefit there goes its complement of violence.

Large harmful projects can be easily avoided by obtaining irrigation from small *nalla* and river dams only. These are also much more cost effective and socially preferable because of the control the local people have over them.

Power generation and use

All industrially-produced articles require generated power for their manufacture, transport and use. Energy is also used by individuals for lighting, refrigeration, air-conditioning, heating and other appliances.

Thermal power stations mainly consume coal, oil or gas, all of which produce carbon dioxide. Carbon dioxide is modifying the climate, as is now acknowledged. Other gases emitted produce acid rain which kills trees and crops up to hundreds of kilometres away, thereby impoverishing the people dependent on them. Coal miners labour under particularly dangerous conditions. The mines cause extensive surface destruction of jungles in the surrounding areas. The burning of coal also produces mountains of ash, which pollutes groundwater, poisoning people and crops. Thermal pollution, produced by the discharge of cooling water into rivers and seas, kills fish and other organisms because they cannot stand the high temperatures.

Nuclear power is by far the most violent source of energy. While it does not produce the type of air pollution that thermal stations do, the radioactivity created is much more harmful and dangerous. All nuclear stations release radioactive substances into the atmosphere in their normal operation. This is most damaging to people living close to the plant. The effects of radioactivity in small doses make their appearance one to three decades later as cancer, mutagenic diseases and birth defects. Studies carried out over many years around power plants show an increase in leukaemia in children.

Even more violent is the production of radioactive wastes in large quantities, for which no safe disposal method has been developed. Since the waste remains dangerously radioactive for thousands of years, this means that violence is being done to numerous future generations. Still worse is the fact that the spent fuel from most reactors can be processed to produce plutonium for bombs which can result in the greatest instantaneous violence known to humanity. The production and the use of all weapons constitute the manufacture and sale of harm.

In the case of a nuclear accident the harm produced can be immense and extensive. Over thirty people died directly in and around Chernobyl; till today, the sale of some farm products is banned in the UK because of contamination from this accident. The radioactivity reached as far as the US, where it has been claimed that thirty to fifty thousand extra deaths occurred in areas where the radioactive cloud deposited its violent material.[2]

All power generation, except some renewable types, pollutes, and nuclear power pollutes absolutely.

If the generation of power is harmful then the use of that power must also be harmful. It is often claimed that electricity is non-polluting, but while it may be so in the place of use, the pollution is invariably produced in the place of generation. The moment a light is switched on, its share of pollution is immediately generated in a power station. For every second that we leave a light on, the carbon dioxide level goes up, non-renewable resources are consumed, and/or radioactive waste is generated.

The use of renewable energy such as biogas, solar, and wind is comparatively harmless, if generated carefully. However, even here there are risks. Large wind generators kill birds which fly into them, so small-scale technology is preferable.

State violence

The continuous encroachment by the state on people's inherent rights produces a corresponding increase in the use of force. This is seen especially in the case of Adivasis and minorities who are driven to desperation because the environment on which they depend is being destroyed. Such people are the immediate victims of "developmental violence". Their demands for more autonomy in order to preserve their habitat, are repressed by force. New laws have been enacted to justify this, although the laws were against constitutional and more fundamental rights and hence were themselves illegal.

Westernization is another form of violence. It diminishes all who retain traditional practices and cultures. Violence is done by the formal educational system which effectively denies certain groups of children the right to enter schools or allows them to enter only to reject them later. But more harm is done by the content of the curriculum which subtly and thoroughly trains children to partake of and contribute to socially-mediated violence. Harm is done when people are forced to flee rural areas because they have no food, water or employment there, and then are compelled to stay in slums from which they are frequently evicted. The use of dangerous pesticides that have been banned in other countries is poisoning everyone in India. The unconcern shown by the government towards the problems of industrial pollution and hazardous wastes is another instance of violence.

Quite often the state gives its citizens no choice in the things they can consume. People in India found themselves eating butter contaminated with Chernobyl radioactivity either directly or in the milk, biscuits and other products made from it. The public was not informed in which items this butter was used and when it was sold. There may be doubt about how much danger is involved with the low amount of radioactivity in the butter but that is not the point. People should know if they are exposed to any extra radioactivity in the environment or in their food, and alternative uncontaminated food should be provided. This would give people the choice whether to accept the risk or not. But the system now forces risks upon the population by deliberately withholding knowledge.

It is becoming increasingly difficult to exercise such choices, since there is rarely an alternative to pesticide-contaminated food, water filthy with industrial wastes and air polluted by vehicle-exhaust fumes.

The content of violence in some commodities

Almost everything we use – whether it is a manufactured article or a service – involves causing harm to others, and one can consider these as having an inherent harm content or volume of violence. It is not possible to quantify the harm content in an object, but studying the whole process, from origin and manufacture to its sale and use, will reveal many interconnections.

Tea

When the British first became addicted to tea they imported so much from China that it resulted in a foreign exchange crisis. They forced our farmers to cultivate the opium poppy instead of food crops, and exported opium to China, even though its sale was officially banned there. Millions of Chinese became addicted to opium merely for the sake of the tea trade. Later, the British cut down Indian jungles for tea plantations. At first, most of what was grown here was exported, but the First World War made sea trade risky. The British then distributed free packets of tea at Calcutta's railway stations and from that time on we have become addicted to it too.

The tea crop in India in 1989–90 was 760 million kilograms, of which 475 million kilograms was used locally, the rest being exported. The total area under tea plantations is about 400,000 hectares, which were previously luxuriant jungles before they were violated for tea. Our thirst for tea is increasing rapidly and so are the exports of tea, requiring further jungles to be destroyed.

Tea-pickers and other workers are miserably paid. But because their remuneration is low, tea is cheap for us and we can afford to drink several cups a day. The low price is achieved at the cost of harm to roughly one million tea-pickers and their dependents. Consider, for instance, how much a cup of tea would cost if the tea-pickers were paid the same salary as the average urban office or industrial worker, which is no more than justice demands. Each worker produces about 700 kilograms a year. Taking an average cost of Rs 50 per kilogram, the annual production per worker is about Rs 35,000. Since the plantation workers are certainly not paid Rs 3000 per month, and there cannot be more than a few thousand other workers in the processing and packaging factories, where does the difference go?

Tea is packed – for export particularly – in plywood tea chests which require further deforestation. Tea purchased in packets requires paper, cloth and aluminium foil, while loose tea does not. Loose tea is sold mainly by small Indian companies, while packaged tea is sold by the big concerns and TNCs: Tata Tea, Liptons, Brooke Bond and others. With every cup of TNC tea, foreign exchange is lost. Most of the tea companies declare dividends of nearly 50 per cent on their shares and these profits must come from every tea drinker's pocket, as well as from the plantation and processing workers.

Not all people can afford tea. For those who cannot, tea then becomes a symbol of luxury for which they must strive. Many, particularly in rural areas, drink tea only to show that they can do so, even though they cannot really afford it. Our example causes harm to them.

The poor use as little tea powder as possible and boil it long and steadily to make it strong, thereby getting more tannin and lead with it. Tea shops often keep tea continuously on the boil which makes the tea particularly bad in this respect. The tannin causes nutritional problems and the lead causes brain damage in children.

With tea goes sugar, much of it from sugar-cane grown in the irrigated fields of Maharashtra. The profits from the sugar industry have given those involved in it the political clout to control irrigation policy. Water is diverted to sugar-cane cultivation instead of being used for food crops and village domestic use. People are forced to buy water and thereby become poorer in order that we may have more sugar. Then there are the migrant cane-cutters who are miserably paid by the factories. Every grain of sugar reduces food production, deprives people of drinking water and adds to corruption.

Neither is milk all that it seems. Much bottled milk is compounded of imported milk powder and butter and partly of milk from the Operation Flood (OF; a 1970s milk-producing scheme) pipeline. The imports keep us dependent on foreign aid and the use of imported equipment for blending and packaging the ingredients. The milk from OF comes partly from crossbred cows which implies a foreign exchange expense. They require a lot of green fodder, so food crop land is diverted to their use, and their special feed requires grains which could be eaten directly by people. The published figures of OF show that they are impoverishing many of their members.

An urban family of five uses more water for making tea twice a day than many rural families get for all their daily needs of washing, cooking and drinking. The lack of water in the villages is a direct consequence of water, as well as other resources, being diverted to the cities. There is also the fuel used for making tea which must add up to a sizeable amount considering the billions of cups of water that need to be boiled.

The food we eat

Food grown using synthetic fertilizers and pesticides has a higher content of harm than food grown organically. Synthetic pesticides are particularly violent because they kill not only the target pests but also millions of other helpful insect species. Some of these pesticides will persist in the environment for hundreds of years. The manufacture of these pesticides is also a dangerous operation, as the Bhopal disaster has shown.

Often the food we eat is grown by big farmers who pay their landless labourers less than subsistence wages. Their use of tractors and other machinery deprives peasants of possible employment.

Processed and packaged foods are much more harmful than unprocessed ones because they contain numerous possibly injurious substances. Buying food produced by TNCs, like Maggi noodles, results in a loss of foreign exchange. Consuming fast foods deprives sellers of traditional foods of their livelihood.

Eleven thousand animals are slaughtered every day in Bombay for meat. Meat, eggs and milk have a higher harm content than vegetables because of the wasteful way in which they are produced. Foods that are transported over long distances are more injurious that those grown and consumed locally because of the fuel and special packaging required for the former.

If all our food and other basic necessities are produced by unnecessarily harmful means, we are forced to commit additional violence in order to survive. In this way the system makes us hardened and callous to causing harm by other means.

Paper

Vast areas of forest are felled for the fibres of trees, shrubs and bamboo, out of which paper is made. Grasses and straws which could be consumed as fodder are used to produce paper. When forest bamboo is used, little is left for artisans who make essential articles like baskets and they have to pay exorbitant prices for what they can get.

Paper production requires much water, so what could be used for food production or even drinking is diverted to mills. And this water doesn't just evaporate, it is highly polluted and then discharged into the rivers from which it is taken. Thousands of people have their drinking and irrigation water polluted and millions of fish and other creatures die because of this. More harm is done if the paper is burned after use as it adds to the carbon-dioxide burden.

If we consider the harm content in the paper used in a novel, magazine and even the paper this book is printed on, we may wonder whether the harm exceeds the benefits received.

The weight of harm can be minimized by recycling as much paper as possible. Every scrap of household waste-paper should be kept separate from other garbage so that it will not be dirtied. This also makes it easier for those who recycle wastes, helping them earn more. Still more harm would be reduced if householders would collect all types of wastes separately and invite collectors to pick them up regularly.

Plastics

Nearly all plastics are made from mineral oil which is rapidly being depleted. Most of the plastics produced in the country are made with foreign technology which means the considerable payment of foreign exchange in technical fees, imported machinery and royalties. Clothes made of synthetic fibre are displacing handloom workers in large numbers.

The production processes invariably pollute the environment. PVC is hazardous to the workers who manufacture it, causing cancer of the liver. Some types of plastic foams used in domestic furniture, insulation and packaging, require the use of CFCs. Some of these foams emit dangerous fumes that kill when they catch fire. Most of the plastics do not decay naturally when thrown away, which is why we see so much accumulating all over the cities and even in the countryside. If they are burnt they produce – among other hazardous substances – dioxin, which is one of the most toxic substances known. Thin plastic bags are floating on the oceans in such large numbers that they are killing fish and other marine life.

Watches

Millions of watches – or their parts – come from Taiwan or South Korea. There, young girls engaged to do this fine manufacturing work are underpaid and mistreated. Their eyesight and health are ruined in a few years. The workers are then discarded as the human wastes of industry. The low price we pay for the watch is the result of violence to them. When the battery runs down it will be thrown out and cause mercury or other heavy-metal pollution so dangerous that Sweden has a law that all used batteries have to be returned to the dealer before a new one is purchased.

Besides, the rich who own watches demand perfect punctuality from the poor who do not, a form of oppression in itself. Finally, we come to the question of time itself: the appropriateness of linear time so closely connected with the linear "progressive development" of the west against the cyclical time of nature and our heritage, which allows of sustainability.

If such small items contain so much harm, how much more must be inherent in large ones like videos and cars?

The export connection

Foreign exchange has to be earned by the export of our products. The current liberalized trade policy permits luxury items (or their components) to be freely imported. The annual imports of components for Maruti cars cost over Rs 150 *crore* while that for video cassettes is about Rs 200 *crore*. Further, much foreign exchange is spent by those who go abroad for holidays and conferences (besides the foreign exchange involved, there is the immense pollution produced by planes and their huge consumption of fossil fuel.) The amount of foreign exchange required is rising rapidly and, therefore, exports of our products have also to rise.

Vegetables, fish and meat, and fruits such as bananas, mangoes and many others are among the items being exported. But when the export of such items is allowed, their local prices shoot up because there is no real surplus available. When the government, in 1987, announced that the export of ordinary rice would be allowed (in spite of the drought), the retail prices of all varieties immediately went up even before the export started. More than 15,000 tons of mangoes plus thousands of litres of mango juice were exported in 1988.

Over one million tons of fish were exported in 1987-8, worth nearly Rs 500 *crore*. This has gone up by more than 130 times in the last 16 years. The price of prawns in Calcutta has risen by more than three times in five years. The per capita fish consumption in India is only 3.1 kilograms compared to the 12.2 kilograms world average. Stopping exports would increase the local consumption by more than a third, but instead, the government is giving added incentives for exporting more fish. Further, traditional fishermen are being impoverished by the government's encouragement of the use of trawlers and other non-traditional fishing vessels in order to increase the catches. Overfishing is endangering many species.

Foods which were formerly cheap enough for the impoverished to buy are now priced out of their budgets. Food exports effectively make people poorer in the sense that they either have less to eat than they need, or that more of their income has to be spent on food. The poor do not use any of the imported raw materials or luxury commodities but they in effect pay for them. The earnings from fish exports are only a little more than what is spent on Marutis and video cassettes.

Yet efforts are being made to export more than Rs 5,000 *crore* worth of agricultural products by the year 2000. To reach that target, incentives such as a subsidy of 20 per cent on the export of fruits and vegetables, and profits from exports, are tax-free.

These, in effect, take money from the poor and hand it over to the exporters. G. K. Leiten states:

> A new series of initiatives has brought the export policy of India quite close to the principles of a market-led policy, as A. S. Ganguly, the chairman of the Anglo-Dutch Hindustan Lever, advised the government in a confidential report: agricultural exports should be freed from "local demand pulls" so that India can establish itself abroad as a reliable supplier.[3]

In addition to exporting, loans are taken from the World Bank and IMF for urban housing, water and sewage, for big dams for irrigation and power, for forestry and for many other projects. These have necessarily to be repaid with interest by more exports later. The infamous conditions under which these are given are that imports be liberalized, that world tenders be called, that foreign experts be involved, and that the government reduces subsidies. While the subsidies on food, electricity, public transport and other items which mainly benefit the marginalized are abolished, the subsidies on exports remain the same or are increased. India's external debt is now Rs 96,000 *crore*, with a debt-service ratio of 38 per cent, implying growing liabilities for future generations.

The volume of violence in imported and smuggled items is considerable, even though it may not register in existing economic accounting systems.

Permissible violence

Gandhi thought that violence against humans and other living beings was permissible when the conditions of the oppressed become intolerable. When ordinary men and women are ruthlessly oppressed and provoked beyond endurance, they are likely to resort to violence. Although regrettable, the violence of those who have been humiliated and brutalized for centuries is not senseless and serves the useful purpose of helping them acquire self-assurance and dignity. In this regard Gandhi believed that landless peasants had a right to seize land.

CONCLUSION

The doctrine of *ahimsa*, though first propounded by the Jains, is thought to have origins in Harappan civilization. This appears to be confirmed by the fact that no weapons have been found in the Harappan excavations. We need to return to principles like these which have survived for such a long time, if justice is to prevail.

Violence produces disharmony in nature, and is a social sin that pollutes humanity. The social sins become pardonable provided we continuously try to reduce the harm we are doing. The reduction of violence becomes essential if we wish to achieve liberation in this world, whatever we may feel about the next one. Greed is perhaps as violent a pollutant as dioxin.

By solely blaming structural oppression for all the injustice today, we can conveniently continue using the benefits of the unjust state, while waiting for "the revolution". But if we admit to a personal responsibility for the existence of oppression, we cannot wait for structures to be changed, since we ourselves are part of the oppressive structures. By reducing our consumption of violence we can make an immediate, positive, individually-small but collectively-powerful contribution to the overthrow of an oppressive system. We can become agents of change by putting into practice that part of a just society for which we are personally responsible.

NOTES AND REFERENCES

1. B. Parekh, "Gandhi's Concept of Ahimsa", *Alternatives*, April 1988. Much of the ancient Indian material and that on Gandhi has been taken from this article.
2. *The Economist*, quoted in "One Deadly Summer", *Anumukti*, April 1988.
3. *Business India*, August 1986, quoted in G. K. Leiten, *The Dutch Multinational Corporations in India*, (New Delhi: Manohar Publications, 1987).

7. Sa Vidya ya Vimuktaye! Knowledge liberates

One of the tools by which the West maintains its power and affluence is its control of knowledge, which is manipulated and selectively disseminated as "information". This process is aided by those in power here. Yet many of us in India are trying to use that same knowledge to improve social conditions, without realizing that there is no way in which knowledge selected for the control of others can liberate us. It is, therefore, necessary to break out of this circle of oppression, a process which involves self-education and a search for truly liberating knowledge.

The policies of modernization employ economic and technological methods that were designed to impoverish us. A system based on wealth and power cannot cater for the basic needs of all. If we accept its knowledge system, we shall only become more deeply entangled in the nets of international oppression.

The system of the West is also self-oppressing. It encourages specialization, with people knowing more and more about less and less. This produces narrow minds which cannot cross the boundaries of their limited fields. The high rewards which experts enjoy only compound their unawareness of their own impotence.

We are subjected to an information monoculture as harmful as the forest monocultures also propagated by the West. The partial view of the world which we are given prevents most people from realizing that they are in effect mental slaves.

There is, besides, an information overload: so much is dumped on us through sleek magazines and scientific journals, through newspapers and television, through computer data bases and conferences, that we are given no time to digest and analyse even a fraction of it. The sheer volume makes it difficult to separate the trivial from that which is vital.

The method whereby the West appropriates knowledge is as subtle as it is self-serving. Neutral data and facts are transmuted into "information", and then selectively diffused as "truth". In the process, knowledge becomes fragmented and incoherent, suitable for sale in the form of commodity.

Many are troubled by the distorting patterns of information emanating from the West. For this process means that those aspects of knowing the world, which cannot be turned into commodities, are simply suppressed, either sinking into oblivion or being shamed into silence. For example, folk wisdom, traditional healing, and even the human instinct to console each other and assuage one another's pain have become the objects of the rapacious intentions of professionals and specialists, those leaders of other people's lives: a landlordism of knowledge.

One of the more baleful side-effects of information-rich Western society is the way it manages to produce vast areas of ignorance out of its very capacity to disseminate its particular forms of knowledge. This is similar to the process that creates new mutations of poverty out of increasing material wealth. The West specializes in the diffusion of redundant and malignant knowledge in, for example, the realm of advertising, publicity and a relentless entertainment industry. All these replace direct experiential awareness of the world with alien implants, which for the most part serve the purpose of selling more and more things to people, which is what "economic growth" is all about.

All this is part of Western education, science, development, which have been held up to us as the most efficient, desirable and just forms of endeavour ever devised by humanity.

Science to the people

Articles regularly appear in the press here about "taking science to the people", crusades which will popularize science and rationality and eliminate superstition. A *Times of India* article states: "Laudable as these efforts are, they are aimed, by and large, at the urban and semi-urban literate and educated readers. . . . Beyond that lie the rural millions, illiterate and certainly uneducated whose life is governed by superstition".[1]

Superstitions are defined as irrational beliefs, not necessarily religious. These are not, of course, restricted to the illiterate rural millions, but are widespread among the urban educated too. Among the superstitions of the latter may be listed the irrational convictions that material wealth brings happiness, that what comes from the West is superior to that which is indigenous, that Western technology and science can furnish all the answers to India's problems, that progress is both benign and inevitable.

Another common error is the assumption that those who have not passed through the formal education system are governed by superstition. This equates illiteracy with irrationality or ignorance.

Most formally-educated persons naïvely imagine that it does not require any reasoning whatsoever to grow crops, to tend cattle, to build huts or to make the thousands of varieties of handicrafts which are so widely appreciated. But rural people learn for living; knowledge is orally transmitted, and communicated through everyday example and practice.

Farmers use tools and their intelligence to transform land, water and solar energy into food for our sustenance, as well as into raw materials for luxuries, which the clever consume unthinkingly. Without correct decisions on when to plant and transplant, when and how to fertilize, weed, harvest and thresh, agricultural output would be disastrously low. Most rural children who pass through the formal education system are unable to cultivate their fields, and are equally estranged from the age-old artisanal tasks.

It can never be emphasized too strongly that it is the vast knowledge, together with the toil and sweat, of these supposedly ignorant rural millions that is feeding, clothing and maintaining all of us. Those who speak glibly about making a living in industry and commerce fail to observe the extent to which that living is made for them by those who tend the earth and nurture the soil.

The same article couples superstition and tradition, as if all traditional knowledge were superstitious: "When the womenfolk observed for themselves that . . . rice could be cooked more quickly with the new chullas, their minds were ready to absorb the 'science' behind many other facets of their daily chores which were bound by tradition and superstition".

In one sentence, all that has nourished and borne up our ancient civilization for millennia is dismissed as insignificant. The equation of tradition with superstition is irrational and is itself a superstition.

The word science was derived from the Latin word "to know". Its use was later restricted by the West to denote only their specific knowledge of the natural and physical sciences and later, other limited areas of knowledge. This conveniently licensed the West to consider as invalid all other knowledge systems, particularly those of their colonies. The colonizers, by this process, not only occupied our lands but also our minds. And while they have vacated their physical occupation, the foreign rule of our minds extends its dominion daily.

It is essential to distinguish superstition from traditional technology, for which scientific explanations may not be known. Technology can exist without an awareness of the principles on which it is based. All of us use technology for which there is no full explanation in our daily lives. There is no perfect theory for the electron; but that scarcely makes the use of electricity superstitious.

The *Sushruta Samhita* gave seven methods of purifying water, among which were the use of heat and sunlight.[2] Sushruta and his contemporaries certainly did not know of the existence of either bacteria or ultraviolet light. This did not make the disinfection of water by these methods a superstition before microbes were observed.

An ancient herbal medicine was *sarpagandha* (*Rauwolfia serpentina*), which has been used for hundreds of years in India. Only in the last 50 years has Western allopathic medicine discovered the drug reserpine in the plant. Did its use become rational only after it began saving and relieving the lives of millions in the West?

There are many ancient practices for which Western science is only now finding an explanation. Our farmers, for instance, sow seeds at particular phases of the moon. They believe that if so planted, better crops will be produced. This has always been discounted by the educated as superstitious, because the calculable effects of lunar gravity and light on plants are insignificant. Recently, however, it has been found that many of the insects that attack these crops have life cycles in phase with the lunar ones. The crops, which are particularly vulnerable at certain stages of their growth, are therefore dependent on the time at which the insects are at their most destructive.

The article in *The Times of India* also states: "The masses need to be informed about the impressive strides taken by science to change their lifestyle and improve its quality". Western science has certainly taken immense strides to change the lifestyle of us all. But the rationality of a technology that improves our lives by increasing the rate of cancer, genetic malformations and brain damage through pollution is surely questionable. For it is Western science and technology which are responsible for the 100,000 or so chemicals that are irreversibly poisoning the earth.

Possibly the most dangerous superstition in vogue today is the belief that nations can accumulate weapons which could make our planet lifeless, and still expect life to continue indefinitely. The belief that Western science can guide us away from nuclear or environmental disaster is all the more alarming because it is being propagated by those who claim to have a monopoly on rationality.

The empty stomachs of the 300 million below the poverty line cannot be filled with research on the neutrino mass, the origin of the universe, or the usual copycat work that is carried out in our establishment laboratories.

The intensely materialistic society that has arisen out of superstitious faith in Western science has weakened our bonds with the rest of creation. That such a society can resolve the disharmony with nature which it has itself caused, only compounds the error, and leads us into more deeply superstitious behaviour.

In India, most alien technologies have helped the rich minority in villages and cities to add to their advantages over others, while the poor majority are marginalized further.

Wisdom versus science

It is the search for the knowledge that sets us free which we are concerned with here at Maharashtra Prabodhan Seva Mandal (MPSM).[3] MPSM has been studying and collecting the knowledge of the Adivasis (and other rural people) for many years, learning science from – rather than taking it to – the people.

We have our roots in a sustainable civilization that has continued, except for the brief disruption of the last century or so, for over four thousand years. Our traditional wisdom stressed a holistic system, where the interdependence of all animate and inanimate objects made them one whole and entire being. Union was considered superior to division, the undifferentiated to the differentiated. *Vidya*, true knowledge, gave the ability to see the whole as one integrated, inclusive universe. *Avidya*, ignorance, denied that ability. Avidya has now become the cornerstone of the global "knowledge" system.

Part of the knowledge we need is available in our ancient books, some can be obtained from those whose memory of what used to be done is still fresh, still more from those, like the Adivasis, who even today follow the old sustainable methods. The rest has to be worked out by a process of self-education, with the help of others who wish to do the same. Perhaps it is in this search that the people of the West and South may find a common emancipatory goal. Such a task may involve struggle, but it is likely to be more fruitful than disingenuous reassurances to the people of the West that they can have their environmental cake and consume it at the same time.

The formal education system disparages the Adivasis' knowledge of agriculture and their forest environment: it throws away our richest resource, which is what the people know. At a time when even the most ardent proponents of the industrial way of life are recognizing the necessity of conserving resources, we are nevertheless wasting this precious human storehouse with the most reckless prodigality.

An example of the difference between the knowledge of the West and that of the Adivasis illustrates the point: westernized botanists can describe the leaves, flowers, fruits and other parts of plants, and are able to tell you the family, genus and species to which they belong. But put them in a forest and they very probably will not be able to identify most of the plants, unless they examine the flowers and other parts. But an Adivasi child can identify most of them, seen from afar, even when they are bare of leaves and flowers. The Adivsi reads trees as a good reader reads words: in their entirety, at a glance. The botanist, on the other hand, reads trees as a neo-literate reads words, letter by letter. The botanist deciphers trees but cannot read them fluently. The Adivasi can also list the uses of each plant, but the botanist will refer you to an economic botanist or chemist for that. Our education system validates only the botanist's approach, ignoring or suppressing that of the Adivasi.

By placing sufficient value on their traditional knowledge and reinforcing it, the Adivasis will gain the confidence to confront the threats that come to them from outside – the assault on the forest by the alien values of the market economy. This type of knowledge can be self-propagating, without the need for elaborate institutions. It can be diffused by traditional oral means of communication, rather than by television or printed matter.

Rediscovering old wisdom and attaining new, does not need expensive laboratories. As Gandhi said: every man must be his own scientist, and every village a science academy. They may not be able to contribute to pure or "leading-edge" science. Such science, requiring enormous funds, can only be supported by an unjust system of appropriation.

Knowledge of nature's gifts comes neither naturally nor through formal studies: a botanist or a forester confined to a jungle would not survive for more than a few days.

Most of our technology still in use was developed by those who, today, would be termed non-scientific people. The thousands of varieties of paddy that are grown were carefully selected by farmers to suit the climate, soil and other local conditions. Dozens of varieties of mangoes were developed for their delicious taste and size and differing uses. The Bharwads and Rabaris were, and remain, experts in breeding cattle.

When one considers the vast range of species of plants and their multiple uses, it can be seen that a tremendous amount of directed research and development has been undertaken by millions of ordinary people over millennia, resulting in a rich fund of knowledge about renewable products. Taking medicinal uses alone, if one multiplies the number of species of plants (more than a thousand) that grow within a few square kilometres of jungle, by the number of parts of a plant (the leaves, flowers, fruits, bark, roots, root bark and others) that are used, by the number of common diseases and other health usages for which each of these are tested (say a hundred, ranging from headaches to fertility control) and by the number of dosages possible, one gets a faint idea of the stupendous amount of research that has been accomplished. If one takes into account the possible synergistic combinations of herbs, this is increased by several orders of magnitude. And this is only for herbal medicines; there are numerous other purposes for which plants are used, such as food, fertilizers, pesticides, dyes, tans, fibres, fuel, timber and other housing material, oil for cooking and lighting, material for mats, baskets, ornaments, and so on.

Proof that such research is still being carried out even by Adivasis lies in the fact that, for instance, Warlis have medicinal uses for plants – such as the Australian tree *dandhavan*, introduced into the area in 1986 – which are not current anywhere else, even among other Adivasi groups. Adivasis are now using the seeds of *dandhavan* for catching fish. It takes about 2 – 3 years to flower and fruit, so the Adivasis' research has been carried out very quickly indeed. A research paper on the effects of *dandhavan* on fish was published in *Environment and Ecology* in 1988. This required a well-equipped laboratory, experimental fish which were killed and thrown away, two qualified scientists who carried out the research and supporting and supervisory staff. It all resulted in an expensive paper whose information will probably never reach those who could use it. The Adivasi research has cost only a little time, and has already diffused so widely that it is difficult to trace who exactly did the research.

With laboratory and clinical tests now proving – to the surprise and embarrassment of the dominant social groups – the efficacy of these remedies, the new science of ethnobotany has been fabricated to serve as a screen for the exploitation of the Adivasis' knowledge. Ethnobotany is an acknowledgment, by scientists as well as industrialists, of the validity of traditional knowledge. But Warlis do not claim intellectual property rights, freely transferring their abundant information to those whose only aim is to profit from its commercialization. The science-technology-industry complex, on the other hand, assumes a right to appropriate that same knowledge, patent it, and then deny people's right to use it freely.

The development experts, the economists and even many ecologists have ignored indigenous knowledge system because of their infatuation with the Western model. But the knowledge of the latter, built up in a mere few hundred years, is already dragging it to the edge of disaster. Anthropologists claim that the Adivasis are primitive, the formal education system devalues and displaces their knowledge, the industrial system destroys forests on which it is based, while ethnobotanists frantically try to steal it before it gets lost so that industrialists who finance them can exploit it. The science of the mainstream, however, fractured as it is into analytical specialities, is not structured to incorporate the study of interconnections necessary for a holistic view. It can make no contribution to sustainability.

Our *rishis* (wise men) also realized that deep understanding could not be obtained unless one led a life detached from material distractions. Westernized scientists like to project themselves as modern ascetics who are wholly and solely devoted to the pursuit of truth; but for too many, science is merely an instrument in their search for fame or power.

Much of the technology that is developed in laboratories, even so-called appropriate technology, has proved a failure when taken into the field, for the simple reason that it has been developed because scientists think that the rural masses need it. Even the term rural masses, used so frequently, ignores the reality of large variations in needs and conditions.

This does not mean that all Western science is bad or that all our traditions are good. We must be more discerning in our response to both. This may require that we slow down our pace, retreat into the forests of the mind, in order to read the signs of the times and to interpret them correctly.

NOTES AND REFERENCES

1. Bal Phondke, "Taking science to the people", *The Times of India*, 3 April 1987.
2. The Sushruta Samhita is supposed to have been written down only in the fourth century AD, though the oral tradition that lies behind it goes back several centuries more.
3. MPSM is a voluntary organization started in 1964 for mainly rural development. In the beginning, it accepted the Western model as a basis for its work. Over the years, it has become apparent that the processes set in train were helping a minority of the poor while damaging the environment and impoverishing many more people than were being helped.

8. The Bhagat and the Allopath

The standard solution to the enormous health problems of India, offered by government and many voluntary organizations alike, is to extend the allopathic system: to have more dispensaries to dole out more drugs, to carry out more immunization programmes, to build bigger hospitals with more elaborate equipment, and to provide the personnel and funds to run these programmes. It is claimed that health for all will be guaranteed if allopathy is made more widely available.

But this system has intrinsic defects which either make it difficult for the poor to gain access to it or often impoverish them further when they do partake in it. The benefits of allopathy are well publicized, but its negative aspects are rarely discussed. At the same time, *bhagats* are regularly maligned and the positive aspects of traditional systems ignored.

ALLOPATHY

The claims

We have been taught by its practitioners, by the formal education system, by the media, and especially by big pharmaceutical manufacturers, that the allopathic system has been the main factor in the reduction of disease. It is argued by proponents of the system that the lengthening life-span upholds its claim to oversee health care the world over. However, the allopathic system, operating in India in a free market, readily lends itself to abuse and fraud; so much so that it is difficult for the people of India to judge what are precisely the positive achievements of allopathy and what are its limitations.

The success of free markets depends upon the continuing growth and expansion of business; this applies no less to pharmaceuticals and medicines than it does to the output of any other commodity. There is, therefore, bound to be a critical relationship between the production of drugs and the growth of illness.

It is true that in the last few decades the life expectancy of peoples all over the world has gone up because many formerly fatal diseases, such as TB, diptheria, malaria, and typhoid are no longer so prevalent or deadly. Ivan Illich furnishes evidence that, while allopathy has provided therapies for some illnesses, the remedies were discovered after the incidence of each disease had reached its peak and started declining. The real reason for the fall was better nutrition that enabled those who contracted a disease to resist it, and public-health improvements in sanitation.[1]

For example, at the end of the last century bubonic plague was rampant in Bombay and regularly killed thousands, as many survivors still recall. It was also common then in Egypt and elsewhere, where they resorted to simple sanitary measures which eliminated the disease-carriers – the rats and fleas – and the disease was, effectively, controlled. In Bombay, however, we had the questionable fortune of the presence of a Dr Haffkine, who was then working on a plague vaccine. Haffkine did not permit the introduction of any sanitary measures here, in spite of requests by other doctors, because this would ruin an opportunity to test his vaccines.[2] Countless citizens, therefore, were sacrificed in order that the practitioner of the science of allopathic medicine could boast of the significant strides he had made.

A programme to control TB was started here in 1976 and deaths from the disease dropped from 10,000 in that year to 7,000 per year now. However, in spite of all the *crores* of rupees spent on TB, the incidence of the disease shows no sign of diminishing. There are still over 13 million people suffering today. In Bombay alone, where free treatment is available, there are 1.6 *lakh* patients. This is understandable because TB germs are normally present in all people here but only become active when a person is weak, primarily due to malnutrition. The eradication of TB depends upon adequate diet for all who go hungry.

Illich further shows that neither the proportion of doctors in a population nor the clinical tools at their disposal nor the number of hospital beds is a causal factor in the changes in overall patterns of disease.

Drugs and their manufacturers

As long as the production of medicines is in the hands of pharmaceutical industries which are based on profit, we must expect their products to care for disease rather than promote health.

Adverse reactions to drugs are usually checked in mice, monkeys and other animals but only in human beings after the drugs have been hastily marketed. Examples are thalidomide and analgin.

One of the painkillers most widely sold in India is analgin. It accounts for Rs 7 *crore* annual sales. There are approximately 200 formulations containing analgin currently available.

This drug may produce a severe loss of white blood cells and increase the tendency to bleed. A major manufacturer of analgin is Hoechst, whose worldwide sales of this drug alone was $75 million in 1984–5. When it was found that it could kill the patient as well as pain, Hoechst launched a propaganda campaign. However, the pressure of activists in Germany compelled the company "voluntarily" to withdraw the drug – but only in that country. Hoechst has openly stated that it will not stop selling the drug anywhere else unless forced to do so. Hoechst also exports to Two-Thirds World other drugs which are not marketed in West Germany.[3]

Arun Bal, a consulting surgeon and an activist of the drugs consumer movement, was summarily sacked from his post at the Dhanvantri Hospital, a small 50–bed institution run by the Bramhan Sahayak Sangh in Bombay. Bal had been actively involved with the campaign for a rational drug policy. He had accumulated a wealth of medical data which challenged the claims made by the promoters of analgin. These data have been cross-checked with reputed pharmacologists in and outside the country, and would undoubtedly affect the sales of this potentially hazardous drug.[4]

Dr Bal states:

> According to Hoechst, approximately 18.3 million daily doses of Novalgin (analgin) are used everyday worldwide. Since most of the developed

countries have either banned or severely restricted the sale of dipyrone (analgin) it is obvious that these doses are mainly used in third world countries. . . . Hoechst has been circulating a booklet titled "Facts on Novalgin" among the doctors in India.[5]

In Thailand human guinea pigs were being used in secret clinical trials of analgin. One of the studies involved 60 children between four and seven years – in spite of international guidelines which state that "children should never be the subjects of research that might equally well be carried out on adults". The reason given for the test was to compare the efficacy of analgin and paracetamol in lowering fever, despite the WHO endorsement of paracetamol as the preferred drug for children. Many scientists were baffled by the need for the tests when paracetamol has a long record of safety. At least two of the trials were designed and sponsored by Hoechst.

It was suspected that the trials were being used by Hoechst to generate data which could be used to promote the drug's use, or as a tactic to delay further restrictions to the drug in that country – a stratagem that Hoechst had used successfully elsewhere.[6]

Such practices are not limited to a few multinationals, but are widespread, sanctioned and promoted by the US government, as shown by the Hatch Bill enacted in the mid-1980s. This law specifically allows US companies to export drugs that have not been approved for use in the US, since this would yield an additional $500 million in foreign exchange.[7] This allows companies to test dangerous formulations on us.

The creation of dependency – such as that set up by the widespread availability of tranquillizers – delights drug manufacturers because it ensures continuous sales of the addictive item. To promote Valium alone, Hoffmann-LaRoche spent $200 million in ten years and commissioned some 200 doctors to produce scientific articles on it.

In addition to addictive drugs, those which require to be taken continuously – for instance medicines for high blood pressure – are also among the pharmaceutical firms' big money-spinners. Sales are further increased by packaged combinations instead of single items. The combinations are medically meaningless and could even be harmful.

Profits are also multiplied by the sale of drugs that actually produce more sickness in the form of dangerous side effects. These require major treatment or even appear as new diseases. A recent report says that one-third of the admissions to hospitals in the US are due to side effects or complications in treatment, or errors in surgery, and are sometimes fatal. One widespread side effect is allergy, caused by aspirin, several antibiotics, steroids and many other drugs. Even antihistamines, used for curing allergies, themselves cause allergies! The system thus has inherent growth promotion, nothing less than the creation, manufacture, distribution and sale of disease.

The cost of a single capsule of one commonly used antibiotic is Rs 6, with four to be taken per day in some cases. This amounts to twice the legal minimum daily wage for rural workers. Since only the affluent few can afford this, the system also determines who is to survive in today's world. We no longer have natural selection at work but a highly unnatural selection – a transnationally determined survival of the richest. The allopathic drugs are not affordable by most of the impoverished, and the new biotech ones will be still further from their reach or will impoverish them yet more. Free herbs are devalued and so are the traditional practitioners who charged just what the patients could afford.

Drugs are evaluated not by the number of people they can cure but by their market value. The *Scientific American*, referring to the development of a new drug using biotechnology, stated that "The stakes are high: . . . the annual market for wound-treatment products can probably already be measured in the billions of dollars". The report, on the other hand, gives no indication of the number of people who would be helped by such a drug.[8]

Drug companies can also corrupt those who control the purchase of drugs in public hospitals. This was the responsibility of Dr R. D. Kulkarni in the J. J. Hospital in Bombay when more than 14 people died because of substandard drugs supplied by Alpana Pharma. Kulkarni was paid Rs 18,000 by Alpana to select their product and he also received Rs 100,000 to Rs 150,000 a year from Hoechst, and Rs 100,000 a year from the Himalaya Drug Co. Justice Lentin, who investigated the case, stated that these payments stopped the moment Kulkarni left the hospital.[9]

Drug manufacturers now claim that biotechnology will produce new drugs that will bring health for all. Yesterday's miracle vaccines and drugs are being suddenly discovered to be ineffective or to have dangerous side effects, thereby creating demand for new wonder products. The effectiveness of the new drugs lies principally in their ability to furnish rising profits. For example, human insulin has been produced by genetically-modified bacteria, and has been available in Britain since 1982. Its developers hoped that human insulin would avoid many of the side effects associated with the traditional, animal insulins, to which some patients are allergic. So far, it has failed to live up to its promise. Many deaths due to low blood-sugar may be occurring among human-insulin users because they do not get the early warning signs that animal insulin gives.[10]

Little research is being done by any of the pharmaceutical TNCs on tropical diseases, for the simple reason that the poor who suffer from them cannot contribute to profits. On the other hand, a huge amount of research has gone into heart and cancer treatment and is going into AIDS because the market in drugs for these diseases is potentially enormous.

Some advocates of allopathy even claim that the availability of numerous drugs now eliminates the need for better nutrition, improving sanitary conditions and even higher incomes for the impoverished!

Doctors

Traditionally, the physician took care of the patient's physical as well as mental health. Listening to the patient's story was itself part of the treatment. Most have now become mere prescription production lines, seeing as many patients as possible but rarely examining them carefully. This profitably converts healthy people into consumers of the doctor's services. Some doctors even refuse to make house-calls since treating patients in clinics results in higher "productivity".

Computers are now being programmed to ask the patient questions, to diagnose the patient's disease and to prescribe, all without the intervention of a physician. This may, to some extent at least, eliminate the effects of doctoral incompetence or a tendency to over-prescribe, but the patient is reduced to the level of an object, a process calculated to swell the ranks of those seeking psychiatric help.

With doctors no longer listening to patients, this much-needed therapy has devolved, in the division of labour, upon the experts in counselling and advising. Many psychiatrists routinely prescribe electro-convulsive therapy to any one who suffers from depression. This treatment can damage the brain, and is rarely used in the West now, but it persists here because of its profitability.

One of the reasons given for the superiority of allopathy is that *bhagats* often, knowingly or unknowingly, give herbs or use rites which have no chemical effect whatsoever. This practice is alleged to constitute fraud. But when doctors use tablets or injections of distilled water as placebos or the rituals of numerous unneeded tests, they are considered experts. City doctors have now discovered that giving an injection of calcium produces an immediate feeling of well-being.[11]

Allopaths, when unable to diagnose diseases, "scientifically" designate them as viral infections or allergies. On allergy, *The Practitioner* says: "From its aetiological basis, to the clinical manifestations and diagnosis as well as treatment, lie many myths; it rivals the virus as a means by which one can explain away puzzling, short-lived ailments and carries the psychological advantage in favour of the doctor that the cause lies in the very being of the patient."[12]

Potent antibiotics are often prescribed when the doctor is unable to diagnose the disease or, even worse, when doctors know that they cannot cure the illness. Surveys show that two-thirds of patients receiving anti-microbial agents get either an incorrect agent or a wrong dose.[13] Such practice contributes to the development of resistant germs. The system creates pathogens for profit.

Some doctors have their own ingenious procedures. The treatment of serious cases of TB requires that drugs be taken regularly for as long as six months, even though patients feel better in a shorter time. Taking advantage of this, doctors have been known deliberately to stop treatment prematurely so that the patient will have a relapse and be compelled to return for further treatment.[14] Each course of therapy may cost Rs 500 for the drugs alone.

The body normally produces pain as a warning of injury or illness. The system has converted pain into a disease that needs treatment. If pain is suppressed by analgesics, it cannot serve its essential purpose as an alarm.

Traditionally, pain has been distinguished from suffering, with the former considered a bodily effect, and the latter mental. Bearing pain then becomes a measure of the inner strength of a person. Pain need not be an obstacle to the fullness of life, unless it fully incapacitates. But most physicians treat pain as an evil in itself. In any case, inner strengths are unprofitable features in a busy market system, and have therefore been considerably weakened in the people of the West, who seek to buy solace for every minor malaise, discomfort or disappointment.

The ambition of most doctors keeps them in cities where patients can pay higher fees than in rural areas. Perhaps this should be considered a good thing for the villagers. However, doctors can scarcely be expected to be ascetics when the whole system is based on avarice.

"Hi-tech" care

Allopathic medicine is seeing an explosive growth in high-technology equipment for diagnosis and treatment. As in all aspects of Western society, complexity is seen as progress, and cost as efficiency or effectiveness.

New technology permits the diagnosis of many genetic diseases. It enables physicians to detect illness and to treat patients early. Some tests pose a number of ethical and legal problems. For instance, the detection of oncogenes or the absence of growth-suppressing anti-oncogenes means that the patient has a high likelihood of getting cancer. When this information is given, it can cause so much anxiety that it leads to further disease.

Even more baleful is the use of tests by employers to detect which employees are more susceptible to disease. This enables employers to save on absenteeism and health benefits; such persons are dismissed or not appointed. Further, insurance companies are seeking to make several tests compulsory, which means that those most likely to fall ill and to need insurance will be denied it. Most tests have high error rates so that people may be further discriminated against.

The tests are naturally limited to those that have been invented; to the equipment available and the pathologist's ability to use it correctly; to the patient's capacity to pay; and to the extent of the doctor's avarice. In this complex machinery of detection, the real disease often goes unnoticed, when a cursory examination of the patient's history and symptoms would give a sure diagnosis.

Cardiac intensive care units require three times more equipment and five times the staff needed for normal hospital care. Some hospitals here refuse to admit patients even when they have a severe attack, unless the charges are paid in advance and in cash (Rs 4,500, in some hospitals in Bombay), which may mean delays of hours or days. The charges for a routinely prescribed heart by-pass operation are Rs 60,000 at a typical hospital here. All must be paid in cash, much of it before admission into the hospital.

A kidney stone crushing machine costs about Rs 2.5 *crore*, with treatment costs of Rs 10,000 to Rs 20,000, whereas most stones can be dissolved with herbal remedies such as *varuna* or changes in diet. The promotion of expensive equipment is often at the cost of more essential items.

Spectacular "hi-tech" therapy convinces people that allopathy is the best source of health care, even though they are excluded by poverty from the medical sanctuaries. Heroic technology tries to keep the rich alive, literally at all costs, no matter what the state of their mental and physical faculties.

Not only are the poor debarred from the supposed benefits of such technology, but it is they who pay for it. The squandering of scarce public resources on expensive equipment prevents funds being used for preventive and basic health care, such as the provision of clean drinking water. And since hospitals earn very little, if any, foreign exchange, the foreign exchange for imported equipment has to come from the impoverished, who have to go without their staple foods which are being exported.

The technical advances in surgery have led to what is probably the worst aspect of the commercialization of disease: the open sale by living people of their own organs. The much-acclaimed development of organ transplants has led to a rise in demand for good organs to replace diseased ones. The supply obtained from people who have just died is inadequate so there is a demand for organs, particularly kidneys, from live donors. Since one cannot imagine rich persons selling their kidneys, these must come from the impoverished who often openly advertize the availability of their kidneys. Recently, it was reported that a person was lured into a hospital for supposed tests but his kidney was removed without permission.

This medical cannibalism, sanctioned and openly promoted, is an indicator of the differential value placed on human life by the contemporary Western world. The barbarism of the system provides "health" for the economically rich in exchange for death for the impoverished. The increasing addiction to technology suppresses moral values since it is a tenet of technologists as well as a compulsion of the economic system that whatever is invented or discovered must be used for maximizing profit.

Public health services

While the urban impoverished are fairly well served by municipal and government institutions, according to allopathic standards, the public health system in the villages is often inadequate or entirely absent. This is hardly surprising since the funds allocated are miserable: 75 per cent of the total funds presently allotted are being diverted to equipment for urban hospitals.

Manglya, an Adivasi, had advanced TB and went to the nearest Public Health Centre (PHC), six kilometres away (bus fare Rs 2.20 there and back). From there he was referred to the nearest district hospital for tests, 30 kilometres away (bus fare Rs 9.80). At his first visit, he could not be X-rayed because of a power failure. At his next visit a few days later, the X-ray machine was out of order. After several more visits the machine was working and power was available simultaneously. He was told to stay in the hospital but he had to make his own arrangements for food. His wife would have had to stay with him, but since she had to work in her fields, he had to return home, untreated.

The public health system does not provide extra food in cases of TB as part of its function. A six months' course of daily injections was prescribed for Manglya. The ampoules were supplied free by the PHC, but the Village Nurse claimed that she was unable to give the injections. As he could not afford the village doctor's fees, Manglya had to go to the PHC for them. Since he had no money for the daily bus fare either, he walked six kilometres each way to the PHC, in spite of his raging fever. Although he appeared to be improving, he could not work because of his weakness and the time spent going for treatment. His wife deserted him, driving him to depression and drink. A neighbour undertook to look after him, but in exchange for all his land.

Village Health Workers get only token training in diagnosis. They are given very few medicines to stock and if they need anything more have to go the PHC, but they can go there only once a week. As a result, the sick may have to wait many days for treatment. A Health Worker gets a miserable salary of about Rs 150 per month and has to attend to several *padas*, often half a kilometre apart.

The PHCs rarely have in stock even common drugs. Doctors are then able to tell patients that the medicine is available only later, for private treatment. Public hospitals, in rural areas at least, are badly equipped and under-staffed. It is now government policy to charge for all services in public hospitals.

Although the Statement on the Indian National Policy declares that "The Constitution of India . . . aims at the elimination of poverty, ignorance and ill health", the actual health policy is mainly concerned with services which are often a source of income to bureaucrats and Big Business. The failure of the government to implement the Hathi – committee recommendations and to reduce the number of drug formulations from over 16,000 to the 200 recommended by WHO is a clear indication of where its interests lie.

In the West, industrial and commercial interests effectively destroyed self-help. This made the introduction of the welfare state an urgent necessity for the impoverished. The dependency which this set up eliminated all memory of earlier methods of self-healing. The dismantling of the welfare state leaves no option but the free market. The wheel has come full circle, with this difference, that there is no longer any choice for those dispossessed of the ability to care for themselves and each other. This is development in action and this is the path we are following.

Mobilizing people to fight for their health rights is not going to be very useful when those rights are limited to allopathic services. Since the allopathic system is a tool for the promotion of social injustice, changing the structures without changing the tools is self-defeating.

The allopathic system as a whole

The allopathic system treats the human body as a machine to be maintained and repaired by qualified mechanics in service stations called clinics and hospitals. The system lays sole claim to validity. Only what doctors can diagnose, label and treat are defined diseases, the rest are ignored as malingering or illusion. All who do not use allopathic facilities are treated as deviants, criminals or at the minimum, irresponsible citizens.

This attempted monopolization of health care restricts the liberty of people to treat themselves cheaply and perhaps even more competently. The allopathic system reaches only a fraction of our people now, while the rest still depend on traditional knowledge. By denying the truth of people's knowledge and their ability to look after themselves, the system is effectively destroying their health.

Medicine is a major instrument of social control through the creation of dependency. What can be obtained free or at small cost from the environment is belittled and viewed as valueless, while what is manufactured and has to be paid for is considered superior and effective. When society demands that we have to pay for whatever we use, what is free becomes worthless, even when beyond price.

The loss of the capacity for creative self-help results in a mental inertia, which adds to the malfunctioning of contemporary society, since it looks to others at a distance for solutions that can be found more easily and effectively locally.

When health care is seen as a service provided by health specialists, its success is judged by institutional criteria, that is, by the number of people attending clinics or by the occupancy rate of hospital beds. The higher these figures, the greater the success of the institution is taken to be. Annual reports try to show the numbers increasing every year. If these figures are valid, it means that there are more sick people, or more people who think they are sick. The real criterion should be the number of people who are and remain healthy.

There is no doubt that allopathic medicine has produced some important benefits. The question is whether the innate defects in the system can be reduced or eliminated to the extent required if it is to reach the poor without impoverishing them further. If it cannot, the system remains a preferential option for the rich.

DEVELOPMENT AND DISEASE

The production of disease by the "development" system comes from its impoverishment of people, by denying them sufficient food, water and health care. It is India's boast that it is self-sufficient in food, but a large proportion of the people simply cannot afford to buy adequate nourishment.

The fundamental right to life in our Constitution is only a negative right. It provides for the punishment of those who kill (but only under certain circumstances defined by the controlling class) but places no obligation on the state to take action while people die slowly of hunger or pollution.

As well as overt exploitation, the ceaseless struggle to survive creates diseases of stress. The unemployed, of course, fare worst of all. Industrial, farm and office workers are under constant tension produced by the continuous introduction of new technology which could either make them redundant or force them to acquire difficult skills. The competition fostered at all levels, from nursery school to the boardroom, places children and adults under permanent pressure. This manifests itself in alcoholism, drug abuse and the multiple forms of social breakdown that accompany the Western path of "development".

Common property resources such as forests are being relentlessly devastated for producing consumer items, and for power and irrigation schemes. This is accepted as necessary because the natural resources, renewable or non-renewable, have no intrinsic value for the system except as exploitable raw materials for big industry. The system ignores the fact that this is depriving multitudes, mainly Adivasis, of their means of livelihood and their access to food and herbal medicines. The shortage of firewood prevents what little food is available from being cooked thoroughly thus reducing its nutritive value.

For Adivasis, displacement from their ancestral abodes and their brusque immersion in alien cultures can also destroy their will to live. Over one hundred thousand of them are being displaced by the Narmada scheme alone. The Narmada scheme is a vast dam project in Madhya Pradesh that will drown thousands of hectares of agricultural land and displace thousands of people. There is currently a question mark over its future, even though it was originally supported by the World Bank and other agencies. Similarly, urban overcrowding has drastic psychological effects on rural migrants accustomed to the traditional landscapes of open fields and wide skies.

Malnutrition is being exacerbated by the promotion of processed foods, in which economic value is added with a decrease in nutrients, which makes an adequate diet more expensive. These are being promoted vigorously by TNCs even in the remotest villages. Shopkeepers are encouraged to give biscuits instead of small change. One child died recently because he had become addicted to biscuits and rejected traditional nagli porridge. Many others have been murdered by infant-food manufacturers who claim that their products are better than breast milk. Indeed, there could be no more glaring example of the ousting of natural and costless processes by the invasion of expensive, damaging value-added methods of answering basic need.

Fifty million suffer annually from water-borne or water-related diseases; over four million children are killed by them every year. The destruction of the forests and watersheds, the excessive use of irrigation for cash crops, the urban bias which diverts rural supplies to cities and sheer exploitation are depriving more and more people of water for drinking and hygiene.

The amount of energy spent by women in collecting water and firewood from places far from their homes results in their dissipating a major part of the energy they obtain from the meagre food they eat. Another major problem is the inhalation of smoke while cooking. This smoke has been shown to contain higher concentrations of carcinogens and other toxic chemicals than cigarette smoke or car emissions. Some women have to purchase water at excessive prices, depleting an already insufficient income.

In the cities, the bias towards the rich provides them with assured supplies throughout the day, while in the poorer areas and slums water is usually available in small quantities at odd hours and often contaminated. Insufficient or inefficient drainage and sanitation systems allow polluted, untreated waste water to go into streams or percolate into wells. Both pathological organisms and industrial and agricultural chemical pollutants are transmitted. An increasing problem is that of contamination in the source areas. Factories are pouring their wastes into the Bhatsa river from which much of Bombay's water is obtained. These toxins cannot be removed by the purification systems and hence are surely poisoning the city's population. Untreated wastes are also dumped into the sea with the result that the waters around Bombay are so highly polluted that fish are dangerous to eat. Synthetic pesticides and nitrate fertilizers are becoming a major cause of sickness in rural areas.

Allergies alone have multiplied five to ten times in the last 20 years, affecting about 10 per cent of the population. The principal cause of this epidemic is industrial toxins. Occupational allergies are common where the following are used: solvents, tars, dyes, rubber accelerators and autoxidants, metal salts and enzymes. People working in catering, cleaning, nursing, hair dressing, building and construction, motor servicing, engineering, tanning and electroplating are particularly affected.

Household articles that cause allergies include perfumes, face creams, eye cosmetics, hair dyes and bleaches, nail polish, lipsticks, talcum powder, after-shave lotions, colognes, shampoos, detergents, home insecticidal sprays, artificial food flavours and colours, chewing gum, preservatives, drip-dry or crease-resistant clothes, artificial jewellery, metal zips, buttons, watches, watch straps, rubber gloves, rubber bands, polyurethane foam and even money plants.

The reduction in the atmospheric ozone-layer results in an increase in skin cancer, while our immune systems are damaged. In effect, we buy illness, for ourselves or others, with practically all the manufactured commodities we purchase. Perhaps this is why the advertising industry is so important – it promotes the desirability of everything, and dissimulates the real costs.

Many cosmetics manufactured by TNCs are highly dangerous, as are some traditional ones. Among these can be mentioned skin whiteners which contain mercury and eye shadows which contain heavy metals. Food additives such as preservatives, colouring agents and flavouring agents have been proved to be toxic. These are mainly present in commercial preparations and fast foods.

Other sources of domestic toxins are paints that contain lead and arsenic, and polyurethane foams used in mattresses, furniture upholstery and for sound insulation.

The burning of plastics can release dangerous toxins into the atmosphere. One of these is dioxin, known to cause cancer and other problems.

The reckless use of biotechnology to produce modified micro-organisms for industry and agriculture will predictably result in the production and dissemination of novel pathogens. The strict guidelines on the dispersal of these into the environment are being ignored by researchers in their haste to put their products on the market. All modern technologies have turned out to have unforeseen harmful repercussions, but these were not usually predictable at the time of their introduction. In the case of biotechnology, however, the dangers are especially obvious.

Many other processes manufacture and supply sickness. Since these diseases produce observable symptoms months or years after the person has been exposed to their original cause, they are ignored by the system. Even effects that are immediately noticeable, such as respiratory diseases in urban areas, are externalized by existing accounting-systems.

It does not appear possible to amend or modify the technology, since pollution is an integral part of the production process or product. Scrutinizing all the chemicals merely to find out which are harmful would require so much expenditure that most items would become prohibitive. In 1985, the estimated cost of good laboratory tests for chronic toxicity and carcinogenicity amounted to about $1.3 million for a single chemical, while the time required was more than five years. Replacing toxic with non-toxic ones in products or production processes – if such could be discovered – would also involve colossal costs. On the other hand, the withdrawal of all untested and toxic chemicals and the prohibition of all polluting production-processes would close down a large proportion of modern industry. But since the internal logic of the system looks on all newly-created disease as a business opportunity, all this shows up as enhanced GNP, and that means economic growth.

While improvements in sanitation and nutrition are reducing illness, technology is introducing new ones. In the West, the life expectancy of people between the ages of 15 and 45 has stabilized because deaths by accidents, cancers and other new diseases of civilization have halted the decline in mortality[1]. If modern development and high technology are to be credited with the production of life saving drugs and other techniques then, in all fairness, that development and technology should also be debited with the diseases and deaths it produces. What would the balance sheet then show?

The optimism of the WHO, with its ambition of Health for All by AD 2000, will remain unrealizable, while the system it serves, and its subsystems, are actively inventing and promoting illness.

HOLISTIC DEVELOPMENT AND HEALTH

Our health is affected by and should affect everything else that we do: agriculture, education, industry, etc, by the whole socio-economic and environmental situation in which we live. Health, therefore, cannot be looked at independently from the rest of society.

We cannot assume that allopathy is the main path to follow while "alternative therapies" are merely inferior options. The rejection of allopathy – where valid alternatives exist – can also be seen as a fight against oppression.

The ancient term *ayurveda* is derived from *ayur*, meaning "life" and *veda* meaning "knowledge". The Constitution of the WHO defines health as "a state of complete physical, mental and social well-being and not merely the absence of disease and infirmity". These definitions go far beyond the curing and prevention of disease.

Preventive health care should cover all aspects, physical and mental, which affect human beings as individuals and communities, and which contribute to preserving and maintaining health. These range from social conditions that lead to good nutrition, to family and community support, to the elimination of workplace hazards and to the prevention of toxins entering the environment.

Health can only be maintained if poverty and its resulting malnutrition are seen as one of the primary causes of disease. To improve nutrition, natural forests have to be regenerated to provide wild foods, fuel for cooking, manure for crops, herbs for medicines and grazing for cattle.

The provision of clean and sufficient water and adequate sanitation are more important than the provision of ORT and other therapies for curing the diseases resulting from the lack of water and sanitation.

Natural defences

Maintaining and supporting the natural immune system, which is weakened by malnutrition, overwork, pollution and other factors, should be given its due importance. Recent studies have shown that family love and nurture play an important role in helping the body's natural defences, particularly in the case of children. It is probable that yoga and other such regimes enhance the body's immunity. The chemicals in garlic, *ashwaganda* and several other herbs have been proved to boost the immune system. However, the effectiveness of garlic is lost when processed into a drug. Many of the cures of faith healers may arise from a strengthening of the will to resist disease.

Using drugs against ailments that the immune system can cope with actually weakens the body, making it dependent on medicines, and also increasing the chances that resistant germs will develop. Further, allopathic drugs are usually not selective, damaging many healthy as well as diseased cells.

One reason for the failure of the immune system to combat disease today could be its extensive overloading, not only by pollutants and micro-organisms, but also by medical interventions such as organ transplants, blood transfusions, serum immunizations and many others. The promotion of baby foods, besides killing directly, deprives millions of children of the natural immunity normally transferred from mother to child in breast milk.

Nutrition

The importance of adequate nutrition for physical and mental growth of children and for the maintenance of health is too well known to need repetition. Undernourishment not only reduces mental and physical capacities of people but also makes them more susceptible to disease.

The fact that the per capita availability of food in the country appears to be sufficient should not lull us into believing that all people eat enough. The lack of purchasing power of the impoverished effectively prevents them from getting adequate nourishment. With over 300 million people below the poverty line, we probably have many more than this figure suffering from malnutrition.

It is important to fight for the positive right to life, from which follows the positive right to food. While immediately this could take the form of adequate ration shops and food entitlements, in the long term it requires major changes in the system: land to grow food, or employment which gives sufficient income to purchase food. This in turn requires the implementation of land-ceiling acts and a proper industrialization policy.

Nutrition can be improved by the proper use of natural foods based on *ayurvedic* principles. While nutrition can also be enhanced by increasing the quantity of food taken in, there are a number of unhealthy practices which reduce the value of what is already available. Some of these practices are:

- the use of polished rice;
- the derogation of coarse grains (which are easier to grow as well as more nutritious than rice) by the example of the rich;
- the use of refined sugar instead of *gur*;
- the use of hydrogenated oils;
- the export of any type of vegetable, fruit and cereal;
- the use of good agricultural land for cash crops (including "social" forestry) which are not required for basic necessities;
- the use of hybrid and high yielding varieties of vegetables and cereals which are bred solely for their appearance, or packaging qualities, rather than for their food value;
- the promotion of exotic vegetables which are less nutritious than local ones;
- the production of high-cost vegetables which only the rich can buy;
- the destruction of common property resources from which many people obtained and still need to obtain, free, clean food;
- the derogation of these free foods by the formal education system, by the allopathic system and by the media in general;
- the green revolution which made farmers switch over from pulses to cereals;

- the anti-nutrition factors in natural foods, such as tannins and phenols, which the poor are forced to consume because better foods are no longer available to them;
- contaminated and adulterated material;
- commercial baby-food consumption;
- commercial child-weaning foods; *and*
- commercial processed foods which have a very high cost to nutrition ratio and are also unhealthy to eat.

When milk-powder or reconstituted milk is used for children and mothers, there should be a check to see whether the person is able to digest lactose. A large portion of the world's population is presently unable to digest cow's milk due to lactose intolerance. If it is not being digested, it can cause diarrhoea and so contribute to further malnourishment. This practice also needs to be discouraged because of the dependency it creates on foreign donor agencies, whose main interest is in reducing the bulk of the European milk-lakes and butter-mountains.

Overeating not only results in heart disease and other diseases of the rich but also in reducing the food available to the poor. The rich monopolize the necessities of the poor without themselves benefiting from the injustice.

General nutrition can be obtained by a simple combination of rice, *dals*, a little oil and green leafy vegetables. Instead of buying pills for iron or vitamins, common vegetables like *palak* and *shevga* and fruits like *awala* can supply the requirements at a much lower cost, if not free. The *ayurvedic* system, however, sees nutrition in a much broader sense than the provision of carbohydrates, proteins, vitamins and minerals. Its dietary advice is based on the individual, the season, the type of work and not just on the food's chemical constituents. Some foods can be safely eaten by all at any time. Others are harmful in certain circumstances when people cannot tolerate them. A third category requires special processing and cooking in order to remove their adverse properties. For instance, pulses are not considered merely as sources of proteins as in allopathy, but each pulse has its own special characteristics which could be helpful or harmful, depending on the factors mentioned above. *Mung* is considered light for digestion and *udid* is heavy. Intellectual workers may find *udid* difficult to digest, while it is right for those who do a lot of manual labour. Mung may be eaten safely throughout the year, whereas *udid* is not good in the hot season. In the third category are tubers, such as yams, that require special methods of preparation to eliminate their toxins.

Vegetables can be cultivated or obtained from wild plants available freely in forests and wastelands, many of whose leaves, flowers and fruits are edible. They are also found in cities where herbs like *kante math* and *kaula* are still harvested from roadsides which are not fully paved. Many wild foods are sold in Bombay's markets. But far more can be obtained if gardens and roadsides are allowed to run wild so that an optimum mix of herbs, shrubs and trees grow. Unfortunately, our notions of cleanliness and beauty, which require the extermination of all wild plants and their replacement with exotics of unknown utility, make all these unavailable to people.

Water and sanitation

Water that is biologically polluted can be purified by boiling but this is usually too costly in firewood to be feasible. Chlorine is extensively used but it produces carcinogens in water that contains organic material. The present practice of putting bleaching powder into village wells could be a source of cancer. Iodine is better than chlorine, but it should not be used by pregnant women.

Traditional methods of purifying water, recorded by Sushruta centuries ago, consisted of using the seeds of *nirmali* and other plants or exposing water to sunlight. Unfortunately, many of the plants are no longer readily available. Keeping transparent bottles or plastic bags containing water in sunlight is very effective, though it may not be possible during the monsoon when the need is greatest. However, most industrial pollutants cannot be removed by these methods so it is essential to prevent, by legal or physical means, the dumping of such wastes into water sources.

Mosquitoes can be controlled by eliminating their breeding places, by the use of larvicides derived from, for instance, cashew-nut fruit, *jendu* or *duranti*, or by the use of plant repellents such as *tulsi*. Chemical pesticide sprays should be avoided since they cause allergies or more severe illnesses, and kill predators that normally keep the insects under control. An example from Borneo is instructive. Insecticides used in villages to control malaria vectors also accumulated in cockroaches, most of which are resistant. Geckoes fed on these, became lethargic, and fell prey to cats. The cats died, rats multiplied, and with rats came the threat of epidemic bubonic plague. The army had to parachute cats into the jungle village.[15]

Curative

Gandhi's principle of *swadeshi* applies to medical treatment, with local solutions preferred to more distant ones. Most illnesses are mild and cure themselves, but where therapy is required, priority should be given to enabling people to treat themselves using traditional remedies. The use of their own indigenous knowledge would also enhance their self-image and would constitute a true empowering of people.

Herbals may require more trouble than just purchasing pills, but they are usually available for the plucking, while pills always cost a lot. To promote their use it is necessary to reinforce the valid knowledge people already have and to give them further information where necessary.

Where home remedies are not available or cannot be used, then indigenous therapies should be recommended. There are those like *ayurveda* which have been tested over thousands of years and found to be valid. Home remedies are usually subsets of these therapies which use locally available herbs and can be easily prepared by non-specialist users.

The common edible herb, *loni*, is used for over 20 diseases, among which are those of the liver, kidney, spleen and bladder. The widespread *bhuiawali*, extensively used as a home remedy to treat jaundice, has been proved to be effective against hepatitis viruses. *Brahmi*, from which an allopathic drug for leprosy, asiaticoside, is manufactured, can be used directly for healing lesions. The resin from the *guggul* plant has been used for millennia for healing wounds and to prolong life. The Biblical myrrh comes from a plant of the same genus. This resin has now been found to have chemicals which are antiseptic, antifungal, and anti-inflammatory and so make valuable dressings. If taken internally, it lowers the levels of cholesterol and triglycerides in the blood. Some of its chemicals enhance the immune system.[16]

There is a common plant, *kuhili*, whose seeds contain L-dopa, effective against Parkinson's disease. The seeds are used in some places by *ayurvedic* practitioners for this disease, but more research must be done on how the seeds can be used directly by the patient.

If the use of such plants is to be extended, it is important to preserve their habitats. It is not enough to grow them in herbal gardens; they should be easily available to all when required. Environmental degradation means that many species which have not yet been tested are being irreplaceably wiped out. Losses produced by big dam schemes like the Narmada one are irreparable.

It is necessary to collect, pool, collate and feed back to the people our extensive home remedy pharmacopoeia, to introduce teaching of the diagnosis of common diseases and their available cures, and the cultivation of medicinal plants, in schools as well as through non-formal education, even to "educated" adults.

Most drug TNCs in India are checking every reference in our ancient literature to herbal usage. They have tested many and found them efficacious. An example is *garwal*. The pickled roots of this common herb are consumed by a certain group of Jains who are remarkably free from heart disease. The chemical responsible for this has been identified by a TNC here. They are patenting the drug abroad since this is more profitable. This is how neo-colonialism continues its malign work today. What belongs to the people is taken away and the benefits returned in the form of expensive products. Those who collaborate with TNCs in this are the equivalent of those who helped the British here decades ago.

While some herbal drugs are found to be ineffective when tested in laboratories, this may be because the method of isolating substances from plants and checking each one separately eliminates synergistic action between the chemicals found together naturally. Since the intention of the TNCs is to identify and synthesize the active chemical, and because this process is time consuming and expensive, it is unlikely that they will be able to identify all the chemicals involved or synthesize them at reasonable cost. They are doomed to failure in many of their quests, but the potency of herbs should not be doubted because of this.

Herbs are more easily available to the impoverished than to the economically rich, to the rural than to urban communities, to the Two-Thirds World than to the One-Third World.

Traditional practitioners

Bhagats have played a very important role in the lives of Adivasis for hundreds of years. They served as doctors long before other types of doctor came to help the Adivasis. Much of their knowledge of medicinal herbs is the same as that of *ayurveda*, and it is possible that *ayurveda* derived much from them. One *bhagat*, Gangaram Janu Avare, has written a book on the uses of 355 medicinal plants, with a second volume under preparation.[17] This was published by a well known *ayurvedic* doctor who was convinced of their validity.

Good *bhagats* know much about illnesses and how to diagnose and treat people who have them. They can diagnose diseases better than Village Health Workers, whose training is rudimentary. They know which plants provide medicines for which diseases and how to extract the medicine from the particular part of the plant – the flower, root or bark – and to prepare it, often mixed with other plants, for use. These herbs are available free and so treatment is not expensive as it is with medicines bought in shops.

However, even the best *bhagat* cannot treat all diseases, any more than the best allopathic or *ayurvedic* doctors can. There are *bhagats* who are learned as well as those who cheat, just as there are good and bad doctors. Since they are under the direct control of the people themselves, corrupt or unskilled ones are quickly detected and abandoned.

It may be that people have a low opinion of *bhagats* because there is a second kind with which the medicinal *bhagats* are confused. These are the *bhagats* who have learned only *bhutali vidya*, the ability to recognise a *bhutali*, a sort of witch. Very few enter this field as it is not very honoured.

It is often said that *bhagats* take advantage of people's ignorance and exploit them. Sometimes the ailments are not cured by the remedies which they suggest, and again the rituals which they perform have no connection with the ailments at all. Many illnesses are caused by the mind and others get cured without intervention. The rites of the *bhagats* help in curing the former and give people a satisfaction they need. This contrasts with the mechanized medicine practised by most allopathic doctors.

More important, a *bhagat*, being a person of the village himself, is constrained to treat whoever comes to him. Allopathic doctors, on the other hand, effectively imply "if you can't pay, you can die". Often a gift is given to a *bhagat* only after the patient is completely cured, which may be a year or more later.

Not all the actions of *bhagats* are valid. Diseases which they cannot cure are often ascribed to *devis* (goddesses) and elaborate rituals have been developed to please the *devis* so that the disease will not strike the people. One such rite, described below, was formerly performed when a case of smallpox was detected, but is now an annual affair held by those who have suffered an attack of smallpox to propitiate Baya devi.

The people from the *pada* assembled at Yeshwant Gond's, the *pada* head, late in the evening, to the beat of drums. They sang songs to Baya with the *neem* tree as a recurring theme. In the centre of the room sat Lahu Dongarkar, the chief *bhagat*, with Lakshman Rayat and Sadu Tople, who had also studied *bhagatry*. Before them was an earthen pot on a low wooden stool. Mango leaves were arranged around the neck of the pot and kept in place with a coconut. To the left of the pot was a stone lamp, to the right a small bowl containing water and little rice. In front of the stool, rice grains, a few *supari* and *champa* flowers were scattered and six coconuts placed around. Several bunches of *neem* leaves and *champa* flowers were suspended over the arrangement on the floor.

As the drum beats grew louder and the chanting more intense, the *vara* (spirit) entered into Lakshman and Sadu. Cross-legged, their arms stretched rigidly outward, palms on their knees, then clenched into fists, their faces contorted, they shook and vibrated, panting with the spirit's power. From time to time they snatched at the swaying bunches of leaves and flowers, inhaling their scent, then resuming their frenetic motions. As the drum beats died away and the voices faded, the quivering and trembling subsided and the two came to a quiet rest. Twice, thrice, the *vara* invaded; after each possession a break was called, for lighting up *bidis* and flexing weary muscles.

At about midnight, a round of drinks was served to the men. After another hour, all those attending, men, women and children, were served a vegetarian dinner. A vigil was kept, interspersed with singing, till the first cock crowed. The six coconuts were broken, and the *prasa* distributed to the men who'd kept vigil. And Yeshwant

Gond for one more year was assured of the goodwill of Baya devi, for the health of his family, cattle and crops. If he hadn't had the *jagrun* who can say what harm might befall the rest of the year?

Such rituals are not only occasions for community celebration, but have also served as focal points for social and political mobilizations. In the 1920s an important movement started as a propitiation of Baya devi just outside the Warli area. This was taken over by *bhagats* and other Adivasis, who claimed that Baya devi demanded that all Adivasis give up alcohol, fish and meat, and fight the liquor-sellers and moneylenders. The movement spread over a very large area and had considerable effect. Although tolerated at first by the authorities, it was soon vigorously suppressed, particularly after the Adivasis joined up with the Gandhian movement.[18]

Bhagats play important roles in addition to the medical one. *Bhagats* are living cultural centres, treasure houses and transmitters of the culture of the Adivasis in the form of epics, songs and rituals. These are recited and sung on several occasions. *Bhagats* also act as priests at births, marriages and funerals. Some specialize: there are *bhagats* who play the *dhak* and sing at weddings and *bhagats* who play the *ghangli* and sing the song of Kansari. *Bhagats* run traditional medical and cultural education centres. With the gradual disappearance of the institution of *bhagats* these cultural treasures also are vanishing. There is much work to be done on the significance of epic and traditional songs and music in the life of a community and how this helps the practice of medicine.

Even when diseases are ascribed to goddesses, there are purification ceremonies with special bathing requirements. These certainly contribute to hygiene as well as cure, since *neem*, often used in the rituals, is known to be antiseptic as well as to be a cure for several diseases.

Because *bhagats* are becoming rare, displaced by allopaths and disparaged by the educational system, they are now being brought together to share their information with others. They are teaching other Adivasis how to identify medicinal plants as well as how to prepare the medicines. At the same time, ayurvedic as well as allopathic doctors, are reinforcing the *bhagats'* knowledge, showing them how to diagnose difficult cases, and teaching them about other medicinal herbs and their preparation. Traditional knowledge is also being recorded at the same time.

Some of the herbs which *bhagats* use have now been "authenticated" by allopathic methods of testing. The juice of *neem* leaves is used to wash wounds and for other antiseptic purposes. *Neem* has now been found to have strong anti-bacterial properties. *Bhagats* use *korphad* for burns. *Korphad* has been tested and found to be very effective in healing burns and skin abrasions. It has antiseptic action and stimulates the formation of new tissues. The bark of the *shevga* tree is used for digestive problems. It has been found to contain potent antibiotics against salmonella, shigella and other pathogens. The flowers of a common hibiscus (*Hibiscus rosa-sinensis*) are used as a contraceptive. They produce strong anti-implantation activity in women.[19]

Ayurveda has not fallen into disrepute, as some claim. Rather it was intentionally run down by the British and other Western interests so that they could claim that their medicinal system was superior. As colonized people, we had internalized these claims, but now we should attempt to free ourselves mentally and reclaim our heritage. Not simply because it is our own, but because it is superior to anything allopathy has for maintaining our health, and in many cases, better than allopathic drugs for curing diseases.

Ayurveda is not primarily a curative system, but tries instead to maintain a balance between various aspects of the body. This is done mainly through a moderate and sensible diet, a reasonable amount of rest and meditation, and a stress-free way of life. Body and mind are considered together as producers of disease. Therapy also takes into account the climate, the season, the natural constitution and the habits of the patient. The invalid is studied as an individual without generalizations based on the disease alone. *Ayurveda* does not use a list of drugs to be taken when particular symptoms appear but is a science with its own understanding of how the human body operates. The same herb can act in different – even opposite – ways, depending on the circumstances.

A further advantage of *ayurveda* over home remedies is that it uses combinations of herbs which are more effective synergistically. This increases their bio-availability and reduces the side effects, if any, of the main herbs. Its medicines are cheap since most can be and are compounded by the practitioners themselves.

A common misconception about *ayurveda* is that it does not provide for surgery, yet many kinds of operations, from the removal of tumours to plastic surgery, have been performed for hundreds of years. Extracts of *neem* were used as antiseptics during operations. Many diseases that require surgery in allopathy can be cured without it by *ayurveda*. The latter also has specific medicines for healing wounds and eliminating the scars of operations while the former merely provides for aseptic conditions.

Ayurveda is still flourishing in spite of the constraints under which it operates: lack of sufficient official recognition and funds (only 1.25 per cent of the total health budget is reserved for all traditional systems), the disappearing herbs and, above all, Western-induced bias.

Critics say that *ayurveda* is not a modern, scientific system as allopathy claims to be, that it is irrational because it is not based on pathogens as a cause of most disease. Microbes were mentioned even by ancient *rishis*, though according to *ayurveda*, most germs by themselves are powerless to cause disease unless the body's constitution has been undermined. This is precisely what studies of our immune system are discovering now. Further, allopathy studies parts of the body in isolation from each other and the environment, rather than holistically. The science of *ayurveda* is further ahead in its understanding of human beings than allopathy, which requires disease in order to flourish.

NOTES AND REFERENCES

1. Ivan Illich, *Limits To Medicine* (London: Penguin, 1976).
2. Shiv Visvanathan, "From the Annals of the Laboratory State", *Alternatives*, January 1987.
3. "Dangerous Drugs", *Economic and Political Weekly*, 21 November 1987.
4. "Doctor's Dismissal: Invisible Links", *Economic and Political Weekly*, 31 December 1988.
5. Arun Bal and Anil Pilgaonkar, "Counterfact On Analgin", *Economic and Political Weekly*, 4 March 1989.
6. "Lethal Trials", *New Internationalist*, April 1989.
7. "Open Hatch", *Economic and Political Weekly*, 7 June 1986.
8. "Stimulating Recovery", *Scientific American*, November 1987.
9. "Graft route of a 'concoction'", *The Times of India*, 1 April 1988. See also "Drug Disease Doctor", Calcutta 1989.

10. "Human Insulin Comes Under Scrutiny As Number of Deaths Rises", *New Scientist*, 19 August 1989.
11. Dr M. D'Souza, a doctor working in the Konkan region, personal communication.
12. Editorial, "The Practitioner", 1983, 227, 1215, quoted in *The Hoechst Medical Bulletin*, Bombay, no. 9, 1987.
13. "Antimicrobial Resistance", *The Hoechst Medical Bulletin*, Bombay, no. 6.
14. Dr Ravi D'Souza, personal communication.
15. Ivan Illich, op. cit.
16. Colin Michie, "Pharmaceutical magic from the Magi", *New Scientist*, 23 December 1989.
17. Gangaram Janu Avare, *Adivasinche Paramparagat Upchar* (Nashik: *Gogate Prakashan*, 1986).
18. David Hardiman, *The Coming of the Devi* (Delhi: OUP, 1987).
19. C. K. Atal and B. M. Kapur, "Cultivation and Utilization of Medicinal Plants" Regional Research Laboratory Jammu-Tawi, Council of Scientific and Industrial Research (CSIR) 1982. Also *The Wealth of India*, CSIR, New Delhi, 1950–88.

9. Natural versus Formal Forestry

"We can't see the forests for the simple reason that the trees are in our homes, offices and garbage dumps."

Global concern about deforestation focuses mainly on preserving the "lungs of the planet", the "natural heritage of humankind". What is less frequently mentioned is the way in which the market economy is directly articulated to forest destruction in its creation of a growing demand for ranch-bred meat, glossy magazines, fancy wrapping paper, packing cases, photocopies, rayon textiles, toilet paper, furniture, junk mail, greetings cards, cut flowers, chocolates, luxury housing, pineapples, ice-cream, novels, and advertising copy in newspapers that invites us to spend on more forest destruction.

Counteracting all this requires massive forestation, but this is often done in an inefficient or even harmful manner. In some areas of the world, more sensitive forms of diverse reforestation have been introduced. In India there is still too great a stress on formal and institutionalized forestry. This consists of growing seedlings in plastic bags and planting them in pits of precise dimensions, arranged in geometric patterns and filled with choice fertilized soil. This is frequently carried out on land that is first denuded of all natural vegetation. In contrast, natural forestry allows the forest to regenerate in all its complex variety, just as jungles have grown for millions of years.

FORMAL FORESTRY

An example of formal forestry is being implemented by the Maharashtra Forest Development Corporation in Peth Taluka.

Of the 30,000 hectares of luxuriant natural forest in the taluka, 80% has been given to the Corporation to develop. This consists in slashing and slaughtering all that flourished in the old forests, replacing them with hectares upon hectares of teak, interrupted only by a few patches of *khair*.

Thousands of hectares have been replanted since 1975. Regular weeding is still being done, and even the common *karandi* has been rooted out, together with the other species that occur in what is left of the adjacent natural growth.

The trunks of most of the teak trees are sheathed in red mud by white ants. When white ants attack the bark of a tree from the outside, they first build little tubes of mud within which they move and gradually expand them to cover the whole trunk. The mud in that area happens to be red. This is rarely seen in jungles, and could be a consequence of monoculture. There are few birds to be seen because they have little to feed on, and their absence probably contributes to the white ant problem. The first crop of teak was to have been harvested in 1982, but felling began only in 1985.

This project is financed by the World Bank, which demands that only the most profitable commercial trees be planted because of the need for an economic rate of return.

On a smaller scale, social forestry schemes promote monocultures of *kubabul, gando babul, dandhavan* and other, usually exotic, species.

Formal forestry has its advantages. A limited number of species are easy to manage and are profitable in the short run. Formal forests are designed for industrial use. They certainly impress donor agencies, whereas a good natural forest should look as though it has been untouched by hand.

But the drawbacks of institutionalised forestry outweigh any apparent advantages. The Adivasis who live within the Peth project area have been impoverished, and great violence has been done to millions of individual plants and animals. Something must have gone wrong with the Peth forestry methods, since the trees have taken 40 per cent longer to grow than expected, the calculations presumably having been based on rates of growth of trees in the natural state. The timber yield of monocultures may be initially impressive, but the total biomass output will not reach that of a natural forest.

The government encourages Adivasis and others to maintain nurseries for social forestry projects. This leads to the "export" of their valuable top soil: plantation forestry requires that seeds be sown in plastic bags which are filled with a mixture of high-fertility topsoil and manure. The person who runs the nursery has to provide his or her own soil and manure, which are thus "exported" when the seedling in its bag is taken away for planting. There is no means of replacing this except by importing fertility from another field. The money they earn cannot compensate for the loss of this priceless asset.

NATURAL FORESTRY

The North Konkan region remains one of the better preserved moist deciduous forests in Maharashtra.

Such undisturbed forests help regulate the climate, provide a wide variety of products and nurture a vast range of flora and fauna. Planting a few species of trees does not make a forest – even if there are thousands of plants. Such plantations should not be dignified by the name "forest". The terms "natural forest" or "jungle" evoke impenetrable thickets teeming with myriad life forms.

A tree cannot live in isolation, and jungles cannot fully develop without micro-organisms, insects, birds, animals and other trees. A natural forest has a profusion of species, living in complex webs of interdependence: multiple symbiosis on an immense scale. These relationships are necessary for the survival of all – otherwise they wouldn't be there. It is an active system of cooperation, not rampant, unrestrained competition, as is commonly thought.

Any intact forest shows that great quantities of nutrients have been incorporated in the growing plants. Part of the nitrogen embodied has been produced by lightning discharges which have converted nitrogen gas into compounds, which are then washed down by rain. In addition, there are dozens of nitrogen-fixing species, ranging from tiny herbs to gigantic trees. These, with the symbiotic bacteria in their root nodules, are the fertiliser factories of the forests. Some of the nutrients may have come from deep underground when the roots of trees reach such sources. The accumulated material is continuously recycled by plants and animals. In monocultures or plantations with few nitrogen-fixing trees, adding synthetic nitrogen becomes a necessity.

Plants in natural forests complement each other; some attract insects to their own leaves, thereby sparing others, while others have a smell that repels insects. Plants control the density of growth in their neighbourhood by their shade. Deciduous trees drop their leaves, giving plants in the undergrowth access to sunlight. Others release chemicals in the soil which limit growth nearby. Plants warn one another when grazed or lopped. The tree being attacked emits chemicals which are detected by its neighbours. These, within a few minutes, produce extra tannin in their leaves which makes them less palatable to animals. This deters the cattle from taking too much from a particular area, and discourages other cattle from browsing the same trees too soon.

With the exception of some exotic species like *gandhara* and *congress gavat*, it may be that all plants despised as weeds are actually assisting other plants rather than robbing them of nutrients.

The precise role of every living entity in the jungle still remains a mystery. There are the bacteria, fungi and other micro-organisms that process vegetable and animal dead matter, turning it into nutrients that can be absorbed directly by the roots of plants. Insects, too, are vital for the pollination of plants. They control other species and provide food for creatures further up in the food chain. One of their major tasks is the rapid recycling of vegetable matter that has served the plant's purpose. As compensation for their services, they claim a little of the vegetable growth for their sustenance, though many see this as a pestilential activity.

It is commonly believed that all caterpillars are pests to be squashed or sprayed as soon as they appear. There is a particularly nasty-looking hairy specimen which gives a terrible itch when touched. This creature appears towards the end of the monsoon, destroying many plants and as there are millions voraciously chewing away, it seems at first to be causing havoc.

But the caterpillar attacks only plants which come up in the monsoon, particularly annual climbing ones. These grow rapidly and, within a few months, completely smother even tall trees. The caterpillars emerge only after these climbers have produced their flowers and fruit and have sown their seeds for the next season. After this, the climbers have no further use for their leaves. With perfect hindsight we can see that without this precise timing the whole system would suffer. If the caterpillars did not wait for the climbers to fruit, these would have died out long ago. On the other hand, if the climbers had died out, so would the caterpillars. If the climbers were not eaten, they would inhibit the growth of supporting plants.

The caterpillars also convert vegetable matter into fertilizer which can be absorbed quickly by other trees – those which supported the climbers as well as the next set of plants in the annual cycle. The caterpillars are in effect tending the natural forest.

Insects also care for the jungle, inhibiting prolific types of plants from taking over the area. They destroy seedlings growing in unsuitable places, even thinning out big trees where there are too many. For example, *warras* seedlings in deep shade are eaten up by insects while those in the sun a few metres away are unaffected. *Kuda* trees seem to be controlled by caterpillars which eat the top tender shoots when the trees grow too close together. They allow one to develop freely while keeping the rest stunted to fill the lower spaces. Perhaps a tree even invites predators by reducing its toxins or increasing attractants, encouraging insects to feed on them, when its leaves have served their photosynthetic purpose.

When the system is ecologically unbalanced, the work of insects appears to us to be that of pests. Their functions are necessary for the forest to optimize biomass, though they are often viewed as destructive.

Some insects process the dung of other creatures, often carrying it directly to the roots of plants. Ants, earthworms and termites turn over and aerate the soil as well as convert organic matter into manure.

Without flies, one cannot have Paradise Flycatchers. Birds eat insects that have served their other functions. They feed at different levels: wagtails on the ground, golden orioles and ioras at the tree tops and many others in between; sharp-eyed drongos, bee-eaters and dancing fantail flycatchers snatch insects on the wing. This ensures that all zones are covered, giving each species a sector in which to operate without competition. But if there is even a small gap in their food supply, the birds cannot survive – their year-round requirements are assured only by a wide variety of plants.

Reciprocally, plants require a range of birds to pollinate flowers, eat fruit and excrete seeds and fertilizing dung. Birds are essential for the dispersal and survival of many species.

Plants, micro-organisms, and animals all play important roles in the delicately balanced structures of a multiplicity of interlocking food chains. It is well known that when luxuriant jungles are cleared to plant crops, the fertility drops rapidly. It is the swift and unceasing recycling of food, with nothing wasted, that allows natural forests to produce so abundantly.

The jungle has multiple layers of plants as well as many "crops" a year, often overlapping, but all fitting efficiently into tight ecological niches. On the surface of soils and rocks are mosses and lichens; there are many herbs and grasses some only a few centimetres high, covered by larger herbs and shrubs; higher still the big trees under which smaller ones often thrive. On the shrubs and trees are orchids and climbers at every level. In forests with closed canopies, the lower plants are controlled by the shade. But when a large tree dies and falls, it clears space for these to grow again.

At the beginning of the monsoon, the first crops – tuberous plants – shoot up, flower, seed and die, before grass and other herbs can smother them. These grow more slowly because they do not have the stored food of the tuberous ones. Several annual herb, shrub and climber species appear in season. Perennials push their way in spaces that are open to them.

If any strand of the complex web of a natural forest is broken or damaged, the productivity of the system must ultimately drop. The biological diversity of such forests improves their ability to survive, since the abundance of species can handle stresses such as drought, long-term changes in climate, or human intervention.

Yields from natural forests

The timber value that commercial interests see in dead trees is perhaps the least valuable contribution of jungles. The functions of trees, hitherto ignored by economists, are now being seen to be of far greater worth.

One of these is the ability to remove carbon dioxide from the atmosphere. Less photosynthesis occurs in plantations than in jungles.

A natural forest requires biodiversity to optimise the use of energy, soil, air and water resources. These accumulate as biomass in perennial trees and long-lived creatures, or are stored in annual plants and ephemeral insects and micro-organisms, which are recycled rapidly. The long-term (not the immediate) biomass is maximized. Any attempt to hasten the process by planting monocultures may give higher immediate yields, but at great future cost.

Trees are essential for maintaining rainfall. Multiple plant layers absorb solar radiation and reduce that reflected back into the atmosphere. This keeps the atmosphere cool and helps moisture to condense as rain. Water vapour needs some tiny particles on which to condense, even when the temperatures are low. Some of the identified nuclei are bacteria which grow in forests, and fine debris from the forest floor taken up by winds. Trees in plantations may not produce either the bacteria or the required debris. Much of the rain that falls on a forest is transpired back into the atmosphere to fall as rain elsewhere.

Natural forests control the flow of rain water and erosion in watersheds. When rain falls on bare ground, each drop loosens particles, which are easily washed away, with a resultant loss of precious topsoil. The dense growth of trees as well as the multiple layers of plants ensure that very little rain falls directly on to the ground. Roots bind the soil well, while small herbs themselves trap whatever does get loosened. The result is that erosion in natural forests is minimal.

All too often, when trees are planted for reforestation, all shrubs and coppicing stumps are cleared away, so that trees can be planted at precise intervals, in straight lines. Small plants which grow quickly and normally cover the ground in a few days are destroyed. The bare soil is thus at the mercy of the full fury of the rain till the seedlings grow to cover the earth. Digging seedling pits often exposes more soil to erosion.

Many of the trees planted by the forest departments are unsuitable for erosion control. The damage caused by monocultures such as eucalyptus is well recorded.

Forests serve as sponges, absorbing water fast and releasing it slowly. Fallen leaves, humus in the soil and even soil algae improve water retention. This enables streams to run throughout the year even where the monsoon lasts for only a few months. Felling forests sometimes increases the volume of water flowing in the rivers of the watershed, since the quantity that was stored runs off swiftly. This leads to flash floods during the monsoon, dry rivers later, and a lowering of the water table.

Soil is built up over hundreds of years, and will take as long to replace when lost through erosion. The eroded area has a reduced capacity to hold water, which in turn increases the runoff. Erosion also makes it more difficult for plants to grow since they have less soil in which to spread their roots, less nutrients and water.

Natural forests provide many resources such as food, fodder, fertilizers, medicines, fibres, gums, oilseeds, and others. With limited supplies of mineral oil, more chemicals will have to be obtained from plants. It is estimated that the production of oilseeds in our remaining jungles is of the order of 20 million tonnes per year. Harvests should be limited to what they can bear without endangering future yields.

But the most important gift of natural forests is what they can teach us. They can show us why our interference in nature so often makes things worse. They can help us develop a farming system that uses the multiple cropping of nature, all produced without sowing, ploughing, weeding, and other laborious operations. Harvesting is continuous because of the diversity of food available.

What can be done

It is not possible to "plant" a natural forest; that is a contradiction in terms. At most we can help it along.

Where there are forests near barren lands, the natural means of dispersal can transport sufficient seeds for their regeneration. It will suffice to stop further destruction.

Seeds are normally spread by the wind, by flowing water, by birds and other creatures eating the fruit. Others cling to clothes or the skin of animals and are dropped elsewhere. Some are picked up by the feet of animals, while others simply fall or are spread by the explosive opening of pods or capsules. Some seeds germinate better when eaten by animals, for instance those of *babul* when eaten by goats. These creatures are vital although they are considered destructive. We too eat fruit and throw the seeds around.

In some places ancient forests have been wiped out over a large area, or too few species remain. It may be that the normal agents that spread seeds are absent. Here, the availability of seeds is the main constraint and it is necessary to collect the seeds of all plants that formerly grew there. Other areas may have been so degraded that only species that can survive under arid conditions in very little soil may thrive in the first instance.

Seeds can be scattered or sown and left to fend for themselves. This may seem an inefficient way of operating but there are good reasons for it. When seeds are sown directly, they will germinate under the right moisture and temperature conditions. This ensures that those best adapted to the soil and changing climate will grow. Only a few out of thousands may survive, but this is not necessarily inefficient when the labour, cost and low survival rate of seedlings of formal plantations are taken into account.

Even if the seeds are eaten by insects or just decay, they will contribute to soil fertility. If the seedlings are eaten by cattle or goats, they will have contributed to reduced erosion and floods at the beginning of the monsoon, when the damage is usually the greatest. Some protection may be necessary from human destruction since grass cutters, for instance, often chop down all shrubs, herbs and seedlings in order to get pure stands of grass. Cutters often burn old grass for the same purpose. Some plants may be killed by fire, but many native species are unaffected by it.

When young plants of such species as *khair* and *babul* are irrigated they grow quickly, but their pampered roots do not penetrate very deeply. When watering ceases after two or three years the trees quickly die. Such plants under natural conditions grow only a few centimetres in the first year or two, but once their tap roots reach the underlying water table, they shoot up rapidly.

In areas where monocultures have already been planted, herbs, shrubs and other plants should be allowed to grow freely. When the planted trees have been harvested, the natural growth should be allowed to take over.

It is common practice to stop all grazing in areas being reforested. Cattle eat seedlings or trample them. They compact the soil and reduce water percolation. But cattle also recycle nutrients and provide fertilizer. It has been found that cattle keep down the growth of coarse grasses and plants that would otherwise take over the area. Their saliva provides plant growth stimulators which makes the grass grow better when grazed than when cut. It has even been suggested that the hoof depressions allow water to collect and so help seeds in them to grow. Some good fodder grasses, such as *dhangli* and *jhinko samo*, cannot survive even for one year if the land is protected from grazing. This is perhaps because the seeds need to be pressed into the ground by the hooves of grazing animals in order to germinate. In one attempt to promote natural forestry, a climber, *bokudvel*, grew uncontrollably, smothering even tall trees. It would normally be eaten by goats and buffaloes, but because this area was protected from grazing the damage was extensive. Experiments in regenerating patches of a few square meters to a few hectares, have confirmed that – provided there are not too many cattle – native trees can grow well without protection.

Many of our local tree species are unpalatable to cattle and are not browsed by them. They also seem hardy enough to survive occasional trampling. Jungles have a large number of edible shrubs and herbs in addition to trees, and it is possible that cattle prefer to browse on the former. Some trees survive under the protection of thorny or non-edible shrubs.

Thorny trees and shrubs themselves help the process of regeneration; for instance, cactus and *euphorbia* species, *karandi*, *khair* and *babul*. One exotic tree which grows fast and is thorny is *gando babul*. Most farmers dislike it as it is suspected of poisoning cattle. The Indian plant *khejri*, from the same species, should be more acceptable.

Seedlings are eaten because the cattle need fodder. It is absurd to expect animals to stop grazing for a few years while trees grow. Sufficient fodder plants must be provided, with the rest being inedible species.

When exotic species are planted, problems arise. *Kubabul* (*Leucaena leucocephala*), originally introduced in the 19th century, is vigorously browsed by cattle and goats and is destroyed by trampling. *Kubabul* is so attractive to cattle they that have been seen to break down a ten-centimetre diameter tree with one hind leg. However, if protected, it is invasive. Farmers long realized that *kubabul* is not suitable for agroforestry. Their name for it was *ku*, meaning bad; when it was promoted as a wonder tree in the 1970s, its name was changed to *subabul*, *su* meaning good. This tree is now being attacked by a psyllid pest that is destroying them on a vast scale, particularly monocultures, as has happened in the Philippines and Indonesia. In India, *kubabul* has not been planted in large monocultures. The pest, which first appeared here about three years ago, seems to be under control by natural predators.

If grazing is stopped and lopping for stall feeding permitted, farmers will cut off as much as possible from a few trees. This not only causes excessive damage to the tree but also reduces the feed value to the cattle because of the deterrent chemicals produced.

Until the pros and cons of cattle grazing are studied further, it may be advisable to control cattle to some extent, particularly in badly deforested areas, but not to exclude them entirely.

Extreme destruction in the past may make it necessary to compromise on maintaining natural diversity initially and to concentrate on fodder and fuel needs. The sowing of quick growing plants which can provide these in a few months without interfering with the growth of other plants, is a possible solution. For instance, *Sesbania* species (*dhedhar, dhaincha, agasta*) provide good fodder and reasonable fuel, and grow over three metres in one monsoon in the Konkan region. They grow through grass or weeds, fix nitrogen, produce a very light shade and are good for grass. Stems of the annual species can be harvested for fuel at the end of the monsoon.

Our ancient sages retired into the forests in search of wisdom. In desolate monocultures, such sanctuaries may not be found. All we shall find there is a reflection of a deformed vision and the disastrous consequences of our desire to dominate nature.

10. Celebrating Trees, Celebrating Life

Without plants, there would be no other life on earth. Plants – and plants alone – naturally convert the sun's energy into forms that other creatures can use. Plants alone lower the entropy, increase order, on the planet. Plants condition the climate for our survival.

Much of the wealth of India lies in its flora. Nearly every tree, herb, shrub and climber has multiple uses here; there is no such thing as a useless weed or a single-purpose tree. Each plant is a chemical factory of amazing versatility. The number of chemicals that industry makes is miniscule compared to those produced by plants. All these are produced without pollution of any sort, while other services for the maintenance of the planet are being routinely performed.

Over 30 per cent of the medicines used today in the west are obtained from plants, and this number is increasing. Yet most plants have still not been checked for the chemicals they contain or for the uses of those chemicals.

Some idea of the multiple uses of common plants can be gained by looking at just three of them: a tree, a shrub and a herb. These suggest something of the true wealth of common property resources.

THE NEEM TREE

Neem – also called Indian lilac or Margosa tree – is mostly evergreen, except in dry localities, where it becomes leafless for a brief period during February to March. It flowers from January to March in South India and later in the north; restricted flowering occurs in August–September. Fruiting follows a similar pattern.

Habitat

Neem grows wild in the dry forests of the Deccan and is cultivated all over India, especially in arid regions. It thrives as a rule where the maximum shade-temperature is as high as 49 degrees centigrade and rainfall from 45–115 centimetres. It tolerates drought. It grows on nearly all kinds of soils, including clay, moderately saline, alkaline and even nutrient-deficient soil, but does well on black cotton soils. It does not grow in waterlogged soil, saline, in deep, dry sands or laterite outcrops. It thrives better than most other trees on dry, stony, shallow soil with a waterless sub-soil, or in places where there is a hard calcareous or clay pan near the surface.

Cultivation

Seeds should be collected from the tree when fully ripe and sown within two or three weeks as the seeds lose their viability. No seed treatment is required. Direct sowings of fresh seed in the shelter of existing vegetation can be carried out. They germinate in about two weeks. No irrigation is required even in semi-arid areas.

If it is necessary to raise seedlings in nurseries, the seed is covered with a thin layer of soil and sparingly watered. The seedlings can be transplanted during the first rains of the second season, when they are 7–10 centimetres tall and the tap root is 15 centimetres long; if larger seedlings are required they may be kept in the nursery for one more year. The stem and roots may also be pruned.

Neem can also be grown from root suckers, and root- and shoot-cuttings; application of growth-regulators induces root formation.

Although *neem* trees require light, a little shade is desirable during the first season of growth. *Neem* does not grow through grass and needs thorough weeding, especially in dry areas. Seedlings are killed by fire and frost.

The plants establish an extensive root system before aerial growth becomes rapid, which occurs after the first season. Neem trees increase in girth by 2–3 centimetres a year, though more rapid growth is easily attained. The tree begins to bear fruits at about five years and matures in 10 to 15 years.

If *neem* is grown among other crops, it needs careful control, for it may aggressively invade neighbouring fields. In imperfectly drained soils the taproot tends to rot and the trees gradually die off. It withstands pollarding well and coppices freely; early growth from coppice is faster than growth from seedlings.

In India many pests have been recorded on *neem* but none is serious. However, in Niger, a scale insect has attacked *neem* trees severely, resulting in total drying. The effect is so serious that pruning of the affected branches does not avert the problem. The insect is spread by big trucks which brush against branches of infected trees on roadsides and carry the infestation to healthy trees further down the road. Monocultures are even more susceptible in the case of *neem*, and should be avoided. In India, this pest is kept under control by natural predators.

Uses

Having a *neem* tree near one's house is an old tradition in India. It is potentially one of the most valuable of all arid-zone trees, with many commercially exploitable by-products and environmentally beneficial attributes. The pesticidal uses of the *neem* tree have been dealt with in Chapter 5.

Food

The pulp of the fruit is eaten. Leaves are made into soup and curry with other vegetables; the slightly aromatic and bitter taste which they impart to curries is much relished by some people. Green sprouts as well as mature leaves are rich in proteins, calcium, iron and vitamin A. Flowers, generally dried, are eaten either raw or in curries and soups, or as a fried dish in South India. The gum has a high protein content.

Some trees, especially near water courses, naturally exude a sap from the tip of the stem. The fresh sap has a strong smell of fermented liquor with the characteristic odour of the tree and is slightly sweet. It is nutritious and is sometimes made into toddy.

The flowers provide plenty of nectar for bees, though the honey has a slightly bitter flavour. The tree is recommended for planting for honey.

Oil

The shade-dried leaves, on steam distillation, yield a golden yellow essential oil.

The seed consists of a greenish-brown kernel (45 per cent) and a hard shell (55 per cent). The kernel yields a greenish-yellow to brown, acrid, bitter, fixed oil (40–49 per cent of kernel, about 23 per cent of the whole seed), known as oil of Margosa, having a strong, disagreeable odour like garlic.

The pulp of the mature fresh fruits is removed by rubbing against any coarse surface or by mixing them with soil and ash and then trampling them. The seeds after washing and drying are cream-white. The fresh fruit as well as the undried seeds deteriorate rapidly due to fermentation. They should be dried as fast as possible, either in the sun or in driers; frequent agitation is necessary during drying. They should never be stocked in big heaps for more than one day. Drying should not be attempted on wet or damaged fruit; instead they should be depulped and the seeds dried. When dry, the seed-coat is hard and brittle and is easily removed in stone-grinders or decorticators and the shells removed by winnowing.

Properly dried seeds or decorticated kernels can be stored without deterioration for about one year. Storing the seeds for one to three months is necessary for optimum yield of oil. The moisture content of the seeds significantly affects the quality of the oil, too. Although the nature of the bitter principles of neem oil does not depend on environmental conditions, their yield is affected by harvesting immature seeds or by bad storage.

The oil is sometimes extracted by boiling the crushed kernels with water. Oil for medicinal purposes is extracted by heating the kernels in an earthen pot. Oil is extracted generally by pressure (yield 30–40 per cent). Wooden ghanis yield a lighter coloured oil with a greenish tinge; high-pressure expellers yield about 5 per cent more oil which, however, is highly coloured. The residual oil in cake (7 to 9 per cent) can be recovered by solvent extraction.

The oil should be stored in well-closed containers in cool places. The crude oil has to be refined for many uses.

It can be used as an illuminant, though it smokes badly when burning. It is used as a lubricant for machinery.

Soap made from the oil has an unattractive brown colour which still retains some smell. This can be considerably decreased by a second boiling and by salting out the liquid. The soap produces a profuse, though slightly greasy, lather. Neem oil is also mixed with other oils and fats for the production of washing soap.

The oil yields bitter principles and other minor compounds. It can be converted into polyol to substitute a similar petroleum product (polypropylene glycol) used as rocket fuel. Pyrolytic degradation of the oil yields pyronimin, a denaturant for alcohol.

Fuel

Neem is a fast-growing source of fuelwood. The calorific value of its wood is high. The wood makes a good charcoal. The pulp that surrounds the seeds can be used in biogas plants. The shell from the seeds can be used as fuel directly or made into briquettes.

Livestock feed

The twigs and leaves are fed to cattle, goats and camels along with other fodder. They increase the secretion of milk, immediately after parturition. Animals fed on fresh cake were found to be healthy, and their yield of milk good. For palatability, jaggery can be added to the cake. The fresh cake smells bad, but the odour goes on sun-drying. The keeping quality of the cake is good and it is not affected by fungi.

The meal obtained after further processing the cake has an agreeable taste and flavour and is used as cattle or poultry feed.

The fruit pulp is eaten by birds and animals.

Manure

The leaves and twigs serve as a mulch and green manure. The pulp water, obtained during the depulping of fresh or dry fruits, is used for manuring and composting.

Neem cake is an excellent fertilizer, several times richer in plant nutrients than farmyard manure. The alcohol-extracted cake is a better nitrogenous manure than the unextracted cake and by proper processing its value can be increased.

Nitrogen is one of the costliest inputs in crop production when synthetic sources are used. Besides, when urea, for instance, is used for a crop like paddy, up to 40 per cent only is picked up by the crop, the rest being lost by leaching and denitrification. *Neem* cake not only provides nitrogen itself, but it also inhibits the process of denitrification, making more efficient use of urea: 25–50 per cent of urea can be saved by using neem cake or oil. Such use also raises the amount of protein in the grain. Large quantities of fossil fuel could be saved if neem cake were used for coating all the urea applied to rice.

Medicine

Neem is planted in gardens because of its reputation as a purifier of air. It has a profusion of leaves and the high rate of photosynthesis causes it to give out more oxygen during daytime as compared with other trees. Sick persons sleep under a *neem* tree as it is believed that this speeds up their recovery.

Neem has a multitude of medicinal uses, only a few of which are mentioned here. Almost every part of the tree is used for medicine.

Neem extracts have anti-bacterial and anti-viral and anti-diabetic properties. They have been used successfully in cases of stomach worms and ulcers. The stem- and root-bark, gum and young fruits have tonic properties. The root-bark is more active than stem-bark and young fruits. It is beneficial in malarial fever and useful in skin disorder. The gum is soothing and allays irritation.

The leaves aid digestion. Tender leaves along with *kalimirch* (black pepper) are found to be effective against intestinal worms. An aqueous extract (10 per cent) of tender leaves is reported to be effective against some viruses. Leaf-extract yields fractions which markedly delay the clotting-time of blood. A strong decoction of fresh leaves is mildly antiseptic. A hot infusion of leaves is much used in the treatment of swollen glands, bruises and sprains. The essential oil from leaves possesses marked anti-bacterial properties and a mild fungicidal action. A common household remedy is to boil fresh leaves in bath-water to lower body temperature and to cure skin diseases; a decoction of the leaves is also drunk at the same time.

The fruit is purgative, and is used against worms. It is beneficial in urinary diseases and in the treatment of piles, tumours and toothache. The dry fruits are bruised in water and used to treat skin diseases. They are also useful for TB and in eye diseases.

The oil has many therapeutic uses. It possesses antiseptic and anti-fungal activity. The oil taken internally, once or twice a day, is a drug of undoubted value in chronic malarial fevers. It is a common external application for rheumatism, leprosy and sprain, and a useful remedy in some chronic skin diseases and ulcers. The warm oil relieves ear trouble; it cures dental and gum troubles. A few drops of oil taken in *pan* (betel leaf) provides relief in asthma. *Neem* oil extract has spermicidal qualities and tests have shown it can also prevent a fertilised egg from being implanted in the uterus. *Neem* oil has been found to be 100 per cent effective in preventing pregnancy in humans when applied intravaginally before sexual intercourse. It has no side effects. The oil is used in medicated soaps. In the uses of neem, we can see how the allopathic system depends on traditional knowledge of natural products. Preparations of nimbidin, extracted from the oil, are externally applied for skin disorders. Another chemical, nimbin, has been used to bring down fevers. Sodium nimbidinate has potent diuretic properties. A good response was observed when it was administered intramuscularly in congestive heart failure. It reduced blood pressure in experimental animals. Sodium nimbidinate and sodium nimbinate have spermicidal activity.

The dry seeds possess the same medical properties as the oil, but they need to be bruised and mixed with water, or some other liquid before they can be applied to the skin or ulcers. The sap is useful in skin diseases, TB and general debility. The bark is used for malaria fever. The bark, with the addition of a little coriander and ginger powder or bruised cloves or cinnamon powder, is said to be superior to quinine. Tender twigs of neem are used to clean teeth, particularly in cases of pyorrhoea.

Veterinary

The paste of leaves is useful in ulceration of cow-pox. An aqueous extract of tender leaves possesses anti-viral properties. An ointment prepared from neem seed oil, neem bark powder, neem trunk water and arjun bark powder in vaseline is used for primary and secondary healing of animal wounds. It seals and dries the wound.

Forestry

Neem trees can be used for shade, for reforesting bare ravines and checking soil erosion, as windbreaks and along roadsides. They can also be used for mixed cropping.

Neem has successfully reclaimed arid barren lands. It increases the fertility and water-holding capacity of soil. The powerful and extensive roots of *neem* have a unique physiological capacity to extract nutrients from highly leached, sandy soil, which are then restored to the top soil in the litter of fallen leaves and twigs.

Timber
The sapwood is greyish white; the heartwood is red when first exposed, fading to reddish brown. The wood is moderately heavy and medium-textured. It is durable and is not usually attacked by insects, even white ants. The timber seasons well even when converted from green logs. Boards cut from green logs should be seasoned in open stacks placed under cover. The wood is very easy to work by hand or on machines, turns on a lathe to a fair finish and lends itself for broad carving, but does not take polish well.

The wood is used for house-building, furniture, carts, axles, yokes, naves and wheel rims, boards and panels, cabinets, packing-cases, ship- and boat-building, helms, oars, oil-mills, carved images, toys, drums and agricultural implements. It is suitable for timber-bridges up to 5 metres span. Pest-proof trunks and chests are made of this wood.

Other Uses
Crop seeds are put in *neem* water in order to make them unattractive, through the bitter taste, to fowls and other birds. *Neem* leaves and oilcake have been used traditionally to improve saline soil.

The bark exudes a clear, bright, amber-coloured gum, called East India Gum, which blackens with age. The gum is soluble in cold water and is not bitter. Trees in dry areas produce the gum very freely. In a wet climate, the gum may be washed away or spoiled before collection. Its adhesive properties are inferior to those of gum arabic. It is used by silk-dyers in the preparation of colours.

Protein fibres have been manufactured from the oilcake. The stem-bark has been successfully tested as a tanning material for goat skins. It also yields a red dye and a fibre which is used for making rope. *Neem* fruit pulp may serve as a carbohydrate-rich base for industrial fermentations. The shell from the seeds can be used for the production of activated carbon, toothpowder and for manufacturing hardboards. The powdered shells are used as fillers in thermosetting-moulding compositions.

Economic importance

A single tree producing 40 kilograms of seeds per year would give about 16 kilograms of oil and 24 kilograms of oilcake. At present prices this gives over Rs 250 per tree per year if the fruit is processed before sale.

It is estimated that there are about 14 million *neem* trees in India, giving a minimum of 400,000 tons of seeds per year. But of this, only about a quarter is used for oil recovery. Ultilising the balance would give an income of over Rs 4.5 million per year.

Plantations of *neem* trees, besides yielding oil, would provide some income for the seed collectors. There would be employment in the oil-crushing and insecticide, fertilizer and soap industries.

THE RUI SHRUB

Rui, also called *ak* or gigantic swallow wort, is a wild shrub, growing all over the plains, in fields and "waste" lands, even in arid regions and cities. It has been traditionally used for medicines and fertilizer. It provides fibre, fuel, pesticides, resins and other chemicals. It is mentioned in the Vedas as in use during sacrificial rites, and is considered sacred by Hindus. They put a garland of flowers on the bachelor-for-life God, Maruti, while Adivasis use the flowers to make the *bashing*, tied on the forehead of the bride and groom. The emperor Akbar is supposed to have been born under an *Ak* bush while *bar* is the liquor prepared from the plant, hence his name.

Description

Rui is a beautiful shrub, 2–4 metres high, with large (10 × 5 centimetres), oval, bluish-green leaves, downy beneath. It has hairy branches. The flowers are pale purple or purplish-white, 3–5 cm in diameter, in bunches, with spreading corolla lobes, appearing practically throughout the year. There is also a beautiful, but rare, white-flowered variety. The fruits are green, paired boat-shaped capsules, about 8–10 centimetres long, bursting when dry and exposing a large number of dark brown, flattened seeds with silky hair attached at one end. The whole plant gives a milky juice.

Children love to see the seeds fly with the wind, which is how it spreads far and wide. A butterfly lays its eggs on the plant; its caterpillars put forth red horns when disturbed. It makes a lovely pale green chrysalis with a band of gold spots.

Uses

Manure and pesticide

Rui leaves have been traditionally used as a manure for paddy in wet lands. The ash has a high potash content. But the fertilizer value of the leaves is not particularly high, being less than 25 per cent of that of normally used green manures. It is possible that *rui*'s value lies in its other properties: its leaves are soft and decay fast, it grows wild even in arid lands, it contains a substance (calotropin) which acts as a systemic insect-repellent and which is taken up by paddy plants, it kills nematodes and other insects in the soil. These functions need further investigation. Extracts of the flowers cause high mortality of the rice weevil.

Rui leaves, incorporated into the soil, control mealy bug attack on paddy. It is used in Andhra Pradesh, when castor is attacked by the red hairy caterpillar. *Rui* extracts act as a sterilizing agent. In Tamil Nadu, its leaves are broken and spread on the field because they attract larvae, which early next morning are collected and destroyed.

The plant seems to be a good candidate for pesticidal use. Only two species of insect can withstand the chemicals in its leaves: the larvae of a swallow-tail butterfly and the highly coloured, eight centimetres long, Ak grasshopper. The fruits are attacked only by a weevil, and the maggots of a fruit fly. All other insects are killed or harmed by its chemicals. When the *ak* grasshopper is attacked by a potential predator, it defends itself by ejecting a spray from a poison gland which kills the predator. The spray contains toxins extracted from the plant.

Reclaiming salt lands

The leaves and stalks have been traditionally used for reclaiming lands covered with saline efflorescence. They are strewn on the ground, covered with soil and then crushed by stamping. The field is then flooded and when the water subsides, the crushing is repeated and the field again flooded. The decomposition of the leaves destroys the salt. Land treated for two years gives good crops.

Medicine

The plant is frequently mentioned by Sushruta. The Arabic name for the drug prepared from it is *ushar*, and the Persian, *khark*.

For elephantiasis, the root bark in the form of a paste or the root or leaves are ground in *kanji* (rice water) and the paste applied on the leg. For scorpion poisoning the root is rubbed in water and the paste spread on the affected area.

The latex is used for removing thorns embedded in flesh. It is used for cleaning wounds, as it kills maggots inside the wound. For dog bites, a paste of the latex with jaggery and oil is applied to the bite. For pain in the thighs and knees the latex and juice of tamarind are rubbed in for three days. For dark patches on the face, a paste of turmeric powder and rui latex is used. Latex, with turmeric powder and *til* oil is effective against skin diseases. For sprained muscles, the latex is first applied to the affected area, and then mustard powder and a rui leaf tied with a tight bandage.

For swellings, leaves are heated and applied to the spot where the swelling occurs. For body ache, hot leaves are placed on the ailing part. For a congested head, mature leaves are tied on the forehead. For earache, *ghee* is applied on mature leaves which are then heated and the juice extracted poured into the ear. For boils and wounds the powder of dried leaves is useful. The twigs are used as toothbrushes.

All parts of the plant are poisonous and care must be taken in the use of any part internally.

The latex is used in combination with that of sabar as a drastic purgative. A tincture of the leaves is use in the treatment of intermittent fevers. Powdered flowers, in small doses, are used in the treatment of colds, coughs, asthma, and indigestion. Powdered root-bark gives relief in dysentery. In small doses it is a diaphoretic and expectorant, and in large doses it is an emetic. For guinea-worms, the flowers are eaten in betel leaf and the *rui* leaf tied on the wound.

For ague, a tender leaf is eaten with betel leaf. For snake poisoning, *rui* leaves are ground in rui latex and formed into capsules to be taken internally, or a paste is made of rui root and taken internally with pepper. For diphtheria, the bark of the tender roots is eaten in betel leaf as *pan*. The flower in jaggery syrup is an appetizer. For migraine, the smoke of burning *rui* twigs or roots are inhaled but it takes two to three days to cure. For cough and asthma, powdered flowers and pepper or jaggery, are eaten.

An overdose causes vomiting. White-flowered *rui* is more effective. There are many other medicinal uses for the plant.

Chemicals
The latex which is present in all parts of the plant contains caoutchouc and resins. It contains a fish poison and an enzyme similar to papain.

The latex was used to a limited extent in the tanning industry for deodorizing, removing hair and imparting a yellow colour to hides. It was made into a paste with flour of the small millet and was used prior to colouring the skin with lac dye. To remove hair from hides, the juice is mixed with salt.

The latex could also be tried out as a waterproofing agent for mud walls.

A plastic material can be obtained from the latex. The thickened and sun-dried sap gives a substance which is thermoplastic in hot water and readily takes any form required, receiving and retaining impressions of seals, ornaments, and so on. It has been made into small cups and other vessels. It may be worth experimenting with the use of the latex, reinforced with suitable vegetable fibres and fillers, for producing low-cost moulded articles for home or educational use, or roofing sheets. For the latter, *rui* fibres themselves, together with latex, could be used directly.

The thermoplastic is a good conductor of electricity, unlike normal *gutta-percha*. It could be checked for possible conductor or semi-conductor use.

Nitric acid converts the latex into a yellow resinous substance. Turpentine dissolves it into a viscid glue which could possibly be used as flypaper, or for trapping cockroaches and other insects, and even rats.

To collect the latex, an open slit is made on the bark of the plant and the juice allowed to flow into a pot. Ten average-sized plants will give a half kilogram of *gutta-percha*. It is not known how often this can be done.

A yellow dye can be obtained from the sap. The stem bark has a colourless wax. The seeds have oil. Resins have been obtained from the powdered bark. A large number of other chemicals are found in the plant.

Fibre

The seeds bear a fine, soft, glossy and resilient floss of cream colour. It is too short to be spun into yarn by itself but it can be spun if mixed with cotton. The floss is very light, and pillows stuffed with it are used for small children and fever patients as they are cooling. They are said to be much cooler than pillows made with *savaar* (silkcotton). The fibre is also used to stuff mattresses and for thread and flannel, carpets, shawls, and handkerchiefs. It could perhaps be used for handmade paper. Selection of long-staple varieties could enable it to be used in a mixture with cotton to give cloth of special properties.

The stem bark yields a fibre which is white, silky, strong and durable. It is superior to cotton and *ambadi* in tensile strength and is used for making fishing nets and lines, twine, strong ropes and bow strings. Extracting the fibre is difficult. Steaming of the stem followed by pressing between wooden rollers, and retting in water with beating with a mallet, have been tried to a limited extent. It can also be extracted by soaking the bark in water for one to two days followed by autoclaving to eliminate incrustations.

Fuel

The wood yields a light charcoal used for gunpowder and fireworks. It can be compressed into pellets. Because it grows easily in arid lands, it could be a possible fuelwood crop. A similar species, *Calotropis procera*, has about five per cent extractable hexane on a dry weight basis. It is possible to obtain one tonne per hectare of hydrocarbon-like materials, equivalent to seven barrels of mineral oil. The plant contains a fermentable sugar which constitutes about eight per cent of the dry weight.

Other uses

Two or three leaves are placed in waterholes to clear water, while its sap is also used for water coagulation in Africa. Its efficiency is not known, neither is anything known about the chemical nature of their coagulating agents.

The plant yields a manna like sugar. A liquor, called *bar*, is prepared from the sap by Adivasis in the Western ghats.

Rui has an exceptionally high evapotranspiration rate, because of which the plant becomes very cool. The temperature of exposed leaves has been measured at 27 degrees centigrade when the surrounding earth was about 40 degrees centigrade. Planting rui around homes should cool the air considerably.

In the old days, leaves were used in place of paper for writing.

Because it grows on arid lands, is not eaten by cattle, and has many uses, it should be seriously considered for plantation on waste lands together with other trees and shrubs. However, its effects on other neighbouring plants needs to be studied first, because of its high evapotranspiration rate, and consequent need for large amounts of water.

Precautions

When growing on or near cultivated fields it may deprive crops of water under drought conditions. It might also prevent other wild plants from growing or thriving.

In case of calotropis poisoning, demulcent and mucilaginous drinks such as milk, rice or gruel should be given, and morphine and atropine administered to allay pain. Skin irritation can be reduced by applying cold water followed by soothing preparations like glycerine or belladonna.

THE TULSI HERB

Tulsi, also known as holy basil or sacred basil, is found throughout India up to 1800 metres above sea level in the Himalayas.

It is grown in and near houses all over the country, and is propagated by seeds. It is susceptible to powdery mildew and root-rot. Two types are cultivated: the green type (*Sri tulsi*) is the most common; the second type (*Krishna tulsi*) bears purple leaves.

Uses

Medicine

The plant is considered an air purifier. Experiments have shown that it releases volatile chemicals that affect bacteria, fungi and human and plant pathogens.

Ancient Indian medical texts advocated the use of *tulsi* as a tonic to increase resistance to disease. Recent studies have shown that one major use could be to enhance bodily resistance against stress. The medicinal properties of *tulsi* tea have now received scientific backing. It has been found useful in warding off and even curing peptic ulcers and other stress-related illnesses. Regular consumption of *tulsi* tea (10 to 20 leaves, mixed with sugar and water) or chewing the fresh leaves relieves stress.

Heart diseases, hypertension, colitis and asthma respond to *tulsi* treatment. It is used in the treatment of viral encephalitis. It increases the survival rate of patients undergoing surgery and reduces post-operative complications. It has anti-inflammatory properties. Water extract of leaves has been found to be as good as or better than allopathic drugs in the treatment and management of ischemic heart diseases.

The juice of the leaves possesses diaphoretic, stimulating and expectorant properties; it is used in catarrh and bronchitis, applied to the skin in ringworm and other cutaneous diseases, and poured into the ear to relieve earache. An infusion of the leaves is used as a stomachic in gastric disorders of children. A decoction of the root is given as a diaphoretic in malarial fevers.

The juice of *tulsi* leaves is very effective in the treatment of coughs, fevers and colds. A decoction of the leaves can cure infections of the upper respiratory tact in a matter of days.

For toothache, fresh leaves are pounded and the juice extracted. Cotton soaked in juice is applied to the aching tooth.

The seeds are mucilaginous and demulcent, and are given in disorders of the genito-urinary system. They contain antistaphylocoagulase which can be extracted with water and alcohol.

Tulsi oil possesses antibacterial properties. It inhibits the *in vitro* growth of *Mycobacterium tuberculosis* and *Micrococcus pyogenes var. aureus*; in antitubercular activity, it has one-tenth the potency of streptomycin and one-fourth that of isoniazid. The oil from the green type is active against *Salmonella typhosa*. Ether and alcohol extracts of leaves are active against *Escherichia coli*. The oil is used as a local anaesthetic.

Tulsi is a common home remedy. For fevers, the juice of 15–20 leaves is boiled in a cup of water, allowed to cool and given to drink. The juice is used with turmeric and honey for cough and asthma. The leaves are used for rheumatism, digestive disorders and worms. For scabies, a paste of *tulsi* leaves, neem leaves, some camphor and coconut oil is applied.

Rats fed on *tulsi* could survive even carbon tetrachloride injections which normally result in fatal liver damage.

Pesticide

The oil has marked insecticidal activity against mosquitoes, though it is not comparable to that of pyrethrum; the repellent action lasts for about two hours. It is planted around a house to repel mosquitoes. Its value in controlling crop and other household pests needs to be investigated.

Sacred usage

Tulsi is the most sacred plant in the Hindu religion; it is consequently found in or near almost every Hindu house throughout India. Hindu poets say that it protects from misfortune and sanctifies and guides to heaven all who cultivate it. The Brahmins hold it sacred to Krishna and Vishnu. The story goes that this plant is the transformed nymph Tulasi, beloved of Krishna, and for this reason it is cultivated, watered and worshipped by all members of the family.

In the *Vrat Kaumudi*, one of the sacred books of the Hindus, a ceremony, called the *tulashi laksha vrat*, is ordered to be performed when a vow is made. This consists in offering a *lakh* of the leaves one by one to Krishna, the performer fasting till the ceremony is complete. *Naivedya*, another ceremonial sacrifice among Hindus, consists in taking a brass dish containing some cooked food, and placing it before the God in a square, previously marked out on the ground with the fingers dipped in water. The worshipper then squats on a low stool, and taking two leaves of the *tulsi* in the right hand, closes his or her eyes with the left, dips the leaves in water, and throws one upon the food, and the other after five peculiar motions of the hand, on the God.

The leaves are used during meditation and *kanyadan* (daughter's marriage). They are also used in funeral ceremonies. The leaves are thrown over balls of cooked rice (called *pind*) made in honour of the dead. A sort of a brush of the leaves is used for sprinkling water at ceremonies in honour of the dead. Vaishnavas wear necklaces made of beads from the roots or stem of this plant.

Traditionally, purifying leaves from the sacred *tulsi* are added to the water used for worship at home and to water offered in temples.

Other uses

The plant is used as a pot-herb; leaves are used as condiment in salads and other foods. The leaves contain ascorbic acid and carotene.

On steam distillation, the leaves yield a bright yellow volatile oil with a pleasant odour, characteristic of the plant, with a suggestion of cloves.

The seeds of the plant give a greenish yellow fixed oil (18 per cent) with good drying properties.

Under favourable circumstances, it grows to a considerable size, and furnishes a woody stem large enough to make beads for the rosaries used by Hindus, on which they count the recitations of their deity's name.

FURTHER REFLECTIONS ON TREES

It was not until the "modern" west was threatened by carbon dioxide that the intimate relationship between trees and human beings was recognized.

But in India, trees have traditionally been accepted as living beings, and generous ones, which respond to those who are wise enough to understand our benign dependency upon them. It is a small step from this to the perception of a tree as "someone" with godly virtues.

The worship of trees is an example of sanctifying the useful. Nothing is more likely to preserve valuable resources than making them holy. The association of trees with divinity is beneficial, since it curbs mindless felling of trees.

It is probably because *tulsi* repels mosquitoes that it has been sanctified. It is no accident that its botanical name was *Ocimum sanctum* (holy basil).

The felling of *wad* (banyan), pipal and other trees belonging to the *Ficus* genus, is prohibited by religious sanction. These trees are so hardy that they sprout on rocks, walls and high rise buildings, with their roots later growing down to reach the soil. Once grown, their fruits attract birds, bats, squirrels and monkeys, which spread the seeds and in turn bring seeds of other plants, "miraculously" regenerating barren places.

The sacred groves of India are repositories of rare species of plants, carefully guarded from the depredations of industrial society by temples or images of gods.

Gods are endowed with characteristics of plants and God-like characteristics are bestowed on plants themselves. A tree is associated with a God whose qualities it is believed to reflect. That tree is then perceived as a tree most dear to that God. The God, therefore, bestows special favours on that tree and it becomes holy.

The dusky and unpretentious *tulsi*, although so valuable, would have remained unprotected had it not become associated with Lord Krishna. There are many stories of the close relationship between *tulsi* and Krishna, and almost all these stories tell us how, in spite of being surrounded by beautiful maidens, Krishna is more impressed by the pure, simple, beauty of the plant. The relationship goes deeper. Krishna destroys evil, and tulsi protects us from the destructive effects of harmful creatures. It is a common home remedy for numerous illnesses.

In Indian mythology, *tulsi* is depicted as a pious, humble woman. The plant is also humble in its appearance, diminutive, with a pure smell. The flowers give a delicate, feminine touch. On the twelfth day of the Hindu month of Karitika (harvest time, October/November) after Diwali, the marriage of tulsi and Vishnu is celebrated. In some regions, the Hindu wedding season starts only after the celebration of this divine marriage. It is the coming together of two pure entities which brings well-being and prosperity to all.

Lord Ganesha is the saviour of the good and destroyer of evil. His favourite colour is red, the colour of blood, the colour of life. Ganesha's preferred plant is the wild *terda* (balsam) with red flowers. The *terda* is considered a cooling plant, while Ganesha is known as an angry God. *Terda* flowers are offered as if to pacify him.

Another favourite plant of Ganesha is *durva*, a common grass, trampled on lawns. *Durva* is also cooling and it is placed on Ganesha's head. What is trodden on by humans is respected by a God. *Durva* has many medicinal uses.

An interesting rite is the "marriage" of *durva* with *wad*. They are a strange pair. The imposing *wad* has the appearance of a contemplating sage, its aerial roots like matted locks and beard. The tiny *durva*, by contrast, clings to the soil. Yet it is not difficult to perceive the bond between them, the *wad* solid, long-lasting with its aerial roots spreading and renewing its strength year after year; *durva*, grass-like, delicate, but with roots that go deep into the earth. Even during drought the roots of *durva* remain succulent and it offers pasture to cattle. This zest for life is the common factor which brings them together. Because of their "ever renewing and ever growing" character, both are considered symbols of fertility and life.

Rui is dear to Hanuman, the monkey-god who helped Rama. Hanuman is known for his strength. *Rui* likewise is a wild, hardy bush which grows anywhere. It is dear to Ganesha also; at his festival, leaves of 21 plants, including *rui,* are offered to him. What is significant is that these gods favour not just the flowers but the leaves also, which otherwise might not be noticed in spite of their immense medicinal value.

Neem is evergreen and offers cool shade. It has magical medicinal properties. On the Hindu New Year Day, a pickle is eaten of *neem* leaves, lemon, raw mango and jaggery. At this time, summer days grow hotter and hotter. *Neem* leaves are cooling. Their bodily cleansing properties prepare people to start the New Year in good health. The evergreen appearance suggests the plant knows some secret rejuvenating recipe, *sanjivani*, the secret of life itself. The taste of the leaves and fruits is bitter, but the appearance is green and fresh. For some people, *neem* is a symbol of Bramha, the creator. In this role, it embodies wisdom.

To usher in the New Year, a *toran* (garland) of mango leaves and marigold flowers adorns doors and entrances. This is the time when the mango leaves are fresh and the tree is in blossom, announcing the arrival of spring. Mango leaves are necessary for any auspicious occasion. They are used in different ways, sometimes as a *toran*, sometimes just a branch hung on the door. The mango is associated with fertility. In the new-year month of Chaitra (March/April), when women come together to worship Gauri, a cooling mango drink is served. Gauri, the consort of Lord Shiva, is the Goddess of fertility, vegetation, creation.

The interconnections between plants, their usefulness, and people's instinctive reverence for them, can tell us much about the nature of our relationship with the earth.

11. Restoring our future

The other day there was a pathetic report in the news about a little girl in Bombay who had never seen a live butterfly. There must be something drastically wrong with the way we have organized our lives – or the way it is organized for us – which has resulted in our exchanging the beauty of butterflies on the wing for a handful of hi-tech trinkets.

The interplay between the internal environment (human appetites) with the external one (the planet) requires to be dismantled and reconstructed in a less damaging and ruinous relationship.

TRADITIONAL RESPECT FOR THE EARTH

Our ancient sages discerned a principle of harmony pervading the entire universe. Each individual forms part of all other life and non-life, one with the earth. This concept requires respect for all that surrounds us, since the individual self merges with the rest of creation. Such a perception can form the basis for a just, sustainable society.

There is the millennia-old belief: "Eko devah sarvabhootheshu tishthati" (One God dwells in all living beings). This leads to the notion of "samadrishti"; "sama" meaning "equal", and "drishti" being "vision" or "sight" – so "samadrishti" means "perceived as equal". A person who possesses samadrishti perceives, experiences and accepts all people as equal.[1]

With this concept of equality, any action that creates or maintains differences is unjust. Distributive justice follows from it. The existence of structures of power and hierarchies is unjust: all must be equally empowered. The non-material requirements of human beings are no less vital. Satisfying basic material necessities for all is only the first step in the search for equity.

All other creatures, too, should be treated as of intrinsic worth and value. The Mahabharata, in fact, recommends that people follow a mode of living which is founded upon a total harmlessness towards all creatures (or in case of actual necessity) upon a minimum of such harm.

On the non-living side, there is the concept of the Pancha Mahabhutas, the five basic elements: prithvi (earth), aap (water), tejas (fire or light), vayu (air), akasa (space). Bhutas are living beings, creatures. Mahabhutas are prior and superior to living beings, since the very existence of the latter is totally dependent on them. "All living beings are born and evolve out of the five Mahabhutas and in death they go back to them." This is why the Mahabhutas occupy a revered place in Indian philosophy.

A stone today is worn out by erosion, its constituents forming soil from which elements are taken up by plants which are, in turn, consumed by animals. The stone today could be part of our bodies tomorrow. And even as we live, and when we die, the elements go back into the soil to be cycled and recycled continuously. Water and air are recycled even faster and spread wider. The water from our bodies can reach the other side of the earth and be incorporated into someone else's. The oxygen that we breathe together with part of our food provides energy and becomes carbon dioxide which, exhaled, can again be picked up by plants anywhere on earth. All of us occupy space today which was occupied by other bodies earlier and will be in the future.

It is not just that "no man is an island" socially, but that every woman or man is or may become a part of anyone else or any other creature on earth. The inanimate elements today are incarnated, made into flesh, tomorrow, and what is incarnated in one living being becomes re-incarnated in a number of others. These processes mock those who seek limits to temporary human identities of caste, nation or race. Today's rulers have in them the elements of yesterday's most abject outcasts.

This is the basis on which so much of the earth's resources become renewable, not only plants and animals but also the common pool of our human flesh and blood.

Those who believe in reincarnation have an even stronger impulse to conserve and protect the environment, for they themselves will literally inherit the earth.

The pollution of the Mahabhutas by industrialized societies, which consider them trivial things to be misused and abused, must necessarily result in violence to ourselves. For when we pollute the Mahabhutas we contaminate ourselves as surely as if we were consuming the pollutant directly. More and more of the thousands of toxic manufactured chemicals that are dumped into our surroundings are being found in our bodies today.

Just what violence industrialism does to the Mahabhutas, by taking the conquest of nature as the fundamental datum for "development", can be seen all over the world. It is even more clear in India, where sustainable civilization had been built on the principles of respect for nature, the use of its creatures and products in a manner to satisfy basic needs, without jeopardising their continued life and growth.

Those sadly struck by "The Enlightenment" and its long-lasting consequences have still to relearn humanity's basic dependence on each individual living and non-living entity in our environment. The conquest of nature was always a deeply ambiguous idea: since human beings are also part of the natural world, who is conquering what and whom? Western science remains ignorant of the deeper reason of nature.

This connects with the ancient Indian principle of the cyclical nature of all creation. This insight is absent from linear projections of "progress" that have come out of the West. Those who have faith in the certainty of eternal recurrences have a perspective of infinite past and future. This offers the individual a sense of security and significance, since the individual sees herself as part of Rita, the Cosmic Order. For the western sensibility, the death of the individual puts a term to her concern for this world. Because she does not feel part of a wider order, she remains isolated and her death extinguishes her being, is Death itself. The belief in the linear movement of life leads to an inner fear and loneliness. This, in turn, creates the desire for faith in progress, which expresses itself primarily in the process of appropriation and accumulation.

In any case, if life continues for the western individual, it is an after-life, in another, heavenly sphere. If the earth undergoes degradation, at least heaven remains inviolate.

But concern for heaven is a minority faith now, even in the West. The consequences of materialism have led to an even greater unconcern for the fate of the earth after the death of the individual. At the same time, in reaction against this, there has been a growing reverence towards the earth because of the realization that this is, after all, the only one we have.

A system which advocates continuous growth is not sustainable even in the short term, whereas a cyclical system can continue indefinitely, provided there is no waste produced. And this is where natural systems are perfect: there is no refuse matter which accumulates and pollutes in nature. Whatever is not required by one creature is utilized as food by others. It is only humans who can and do produce large amounts of persistent waste.

The Buddha came to the conclusion that the problems of the world were the direct result of the desires and greed of human beings. This greed requires much more than basic needs for its satisfaction, with the result that not only is society upset but the environment, too. From this he developed his notion of dharma which was based on the close connections between animate and inanimate nature, the integration of human beings with all the rest of creation. If these connections are perceived by us then we will exercize the required self-regulation so that all human beings can live in harmony with each other as well as nature.

The Jaina doctrine likewise, of jiva-ajiva underlines "the integral view of life, whereby all things, the living and the non-living, are seen as parts of a whole – a single order."[1]

Such thinking is also common among aborigines in every place where old civilizations were not disturbed, destroyed and occupied by invaders. To the Amerindians and Australian Aborigines, every part of the earth is sacred. They believe that they are obliged to use natural resources sparingly, so that future generations will be able to satisfy their basic needs. The desolation of large parts of Africa is a direct consequence of colonial and neo-colonial violations of such fundamental principles.

These principles are remarkable because they were perceived when there was no threat to survival from disappearing natural resources or environmental deterioration. The ability of our ancestors to foresee that these were necessary for the maintenance of life demonstrates the depth of their insight. For it is only within recent decades that western Deep Ecologists claim to have "discovered" some of these basic truths.

Such values, of course, have seldom been accorded universal assent. Rapacious rulers and frequent invaders were more concerned with the accumulation of surplus wealth than with the preservation of their surroundings. Growing populations did result in environmental change. But since the majority of the people were content with basic necessities, such changes never became cataclysmic. But now that "development" has come to mean the legitimation of greed, who can foretell what violence lies in wait for humanity – for the "beneficiaries" no less than the victims?

As long as discussion remains general, it is easy for individuals to feel powerless to influence these great issues, and to turn aside from them. Yet we can practise social justice or rather, to use a traditional term, righteousness, our modern dharma, as individuals, without waiting for the rest of society to change.

CHANGING OUR LIFESTYLE

Western economics claims "consumer sovereignty" as a fundamental "ethical" right. "What I want, I have a right to get", provided only that it does not restrict others' rights to do the same. The criteria people normally use when attempting to satisfy their wants are: "Do I like it?", "Can I afford it?" This implies that the money at the disposal of an individual has been justly earned, that no one has been impoverished or damaged by its acquisition. It further implies that the purchase contemplated has in no way harmed or exploited those who produce it or indeed anyone else. In other words, money itself overrides all other ethical considerations. The whole system is supported by our collusive ignorance of the damage done by the acquisition of our "just rewards", that is, by our money and all it can buy. Such a narrow understanding of self-interest has led, not only to appalling global injustices, but also to the ravaging of the environment. It has taken the latter to bring once more to the notice of the "privileged" the cruel unfairness of the system to their fellow human beings.

When environmental considerations are taken into account, altruism and true self-interest converge and even coincide. Damage to the environment is damage to others is damage to the self. One has to ask: "Is it harmless or beneficial to the environment?", "Does it increase equity?" Housing or hamburgers, transport or toilet paper, every feature of our daily lives, should be subjected to this simple test.

The need for a simple lifestyle logically follows from a holistic vision of the place of human beings in nature. It does not mean a religious asceticism for its own sake, a denial of pleasure. A simple lifestyle can provide a "joyful frugality" which the accumulation of goods and services clearly does not furnish.

For western economics, the lack of a consumable is taken to be loss in utility, a cost to be borne by the deprived person. But those who have voluntarily reduced their consumption find that it increases well-being, liberates from a mere transient quenching of imposed appetites. Reducing consumption then becomes a benefit, a shedding of burdens, personally as well as environmentally.

Development means unfolding – unfolding the full potential of every human being. Human beings have a need for dignity, for love and affection, for care and concern, for the freedom to express their creativity, to control their own destiny, to preserve their own culture, to feel fulfilled, to be educated for life, to know that their life on earth has been worth living. But all this should be and can be within the constraints of universal equity. Indeed, most of these less tangible needs do not require the amassing of wealth, which is why they have so little prominence in existing "development" patterns.

A true flowering of humanity could occur only at a constant, modest, yet secure, level of consumption. It is possible to have better education, health, and quality of life without ever-rising incomes. Perhaps, rather than talking about "nil growth" or "negative growth-rates", which unnecessarily frightens people, we might begin to discuss "positive rates of reduction"; for that would suggest liberation from some of the burdensome, crippling, and indeed, suicidal, consequences of present ways of living.

We can start the process simply by reducing the purchase and use of non-necessities. In a sense, we can make use of the "magic of the marketplace", not only by choosing what we buy, but by choosing what we don't buy.

But even this doesn't address the suppressed question of accumulated intergenerational liabilities. Until these are discharged, what is called aid and charity must be seen as reparations for vast, unacknowledged injuries.

HEALING THE FUTURE

This is not to say that the ideal is possible. One of the problems we, as individuals, have is of reconciling our intention to live a just life with the unrelenting pressures exerted on us by an unjust society. We cannot de-link ourselves completely from the system without becoming *sannyasis*.

We might shift emphasis from personal "success" to personal responsibility. Reducing one's consumption need not wait for politicians' pronouncements or economists' encouragement. The problems have global dimensions but they work through individuals, and this gives each of us the power to exert a decisive influence on the world.

We will be following the path to social justice if we base our actions on the old prayer in the Book of Proverbs: "Give me neither riches nor poverty, but enough for my sustenance."

We need changes in our lifestyle that emphasise voluntary, Buddhist, Christian, Gandhian simplicity. The traditions on whose strength we can draw exist in both eastern and western cultures; it is simply that a possible coming together of these in a creative and holistic alternative has been hindered by the spread of industrial monoculture.

Richard Lannoy, in *The Speaking Tree*, explores the whole range of Indian civilisation and culture. He shows that our traditions make it difficult for us to adapt to the western industrial system. At the end of it he reaches the fundamental and inevitable conclusion that it is not Indians who need to be squeezed into the western mould, but rather that the western system itself is incompatible with life on this planet.[2]

He concludes:

> One feature of universal significance is the importance which Indian civilisation has attached to the simplification and reduction of needs through self-scrutiny. At its most positive a means to reduce social conflict and the dehumanisation inherent in the pursuit of material gain, this kind of humility is rare in western science and technology. It is also the touchstone of our success or failure to reduce tension even within the domain of our personal lives. In an overpopulated world with severely limited resources the current western method – expansion and cultivation of needs – is plainly unrealistic. The wisdom of smallness and the Zero principle, encouragement of small-scale pluralistic activity in community living, a nonviolent ecological perspective, all of which originate in self-scrutiny, are age-old Indian responses to life's dilemmas – the fine flower of crisis.

If the western system of "development" is permitted to endure, of one thing we can be sure: soon, not only will butterflies vanish, but little girls, too.

NOTES AND REFERENCES

1. K. S. Srinivasan, "Tradition Vs. Modernity", The Times of India, 3.10.86.
2. Richard Lannoy, *The Speaking Tree*, (Oxford: Oxford University Press, 1971).

Glossary of Plant Names

adulsa: *Adhatoda zeylanica*
ain: *Terminalia crenulata*
alu: *Dioscorea bulbifera,* yam
ambadi: *Hibiscus cannabinus,* kenaf
arjun: *Terminalia arjuna*
asan: *Bridelia retusa*
ashwagandha: *Withania somnifera*
awala: *Emblica officinalis*
babul: *Acacia nilotica*
bajra: *Pennisetum typhoides,* pearl millet
bakain nim: *Melia azedarach*
beheda: *Terminalia bellirica*
bel: *Aegle marmelos*
bhilangani: *Polygonum glabrum*
bhindi: *Abelmoschus esculentus,* okra
bhuiawali: *Phyllanthus fraternus*
bokudvel: *Combretum ovalifolium*
bor: *Zizyphus mauritania*
brahmi: *Centella asiatica*
chai: *Dioscorea wallichii*
champa: *Michelia champaka*
chaoli: *Vigna unguiculata,* cowpea
congress gavat: *Parthenium hysterophorus*
dalimb: *Punica granatum,* pomegranate
dandhavan: *Acacia auriculiformis*
dhaincha: *Sesbania aculeata*
dhangli: *Brachiaria ramosa*
dhedhar: *Sesbania bispinosa*
diveli: *Jatropha curcás*
dudhkodi: *Wrightia tinctoria*
duranti: *Duranta repens*
durva: *Cynodon dactylon*
erandi: *Ricinus communis,* castor

gandhara: *Lantana camara*
gando babul: *Prosopis juliflora*
garwal: *Coleus forskholii*
ghela: *Xeromphis spinosa*
gowar: *Cyamopsis tetragonoloba*, cluster bean
guggul: *Commiphora mukul*
harbara: *Cicer arietinum*, chickpea
hed: *Adina cordifolia*
hing: *Ferula foetida*, asafoetida
indevi: *Gloriosa superba*
inodi: *Terminalia crenulata*
jaswand: *Hibiscus rosa-sinensis*
jendu: *Calendula oficinalis*, marigold
jhinko samo: *Paspalidum flavidum*
jowar: *Sorghum vulgare*
kachoo: *Colocasia esculenta*
kahandol: *Sterculia urens*
kakad: *Garuga pinnata*
kalimirch: *Piper nigrum*, black pepper
kante math: *Amaranthus spinosus*
karandi: *Carissa congesta*
karanj: *Pongamia pinnata*
karela: *Momordica charantia*
karvi: *Carvia callosa*
kaula: *Smithia sensitiva*
khair: *Acacia catechu*
khejri: *Prosopis cineraria*
khorasni: *Guizotia abyssinica*
kodra: *Paspalum scrobiculatum*
koland: *Ampelocissus latifolia*
korphad: *Aloe barbadensis*
kovli bhaji: *Chlorophytum tuberosum*
kovli-cha-mama: *Urginea indica*
kubabul: *Leucaena leucocephala*
kuda: *Holarrhena antidysenterica*
kuhili: *Mucuna pruriens*
kumba: *Careya arborea*
kusumb: *Schleichera oleosa*
lal ambadi: *Hibiscus sabdariffa*, red sorrel
likandi: *Ixora arborea*
loni: *Portulaca oleracea*
mahua: *Madhuca indica*
math bean: *Vigna aconitifolia*
mundi: *Spheranthus indicus*

mung: *Vigna radiata*
nagli: *Eleusine coracana*
nilgiri: *Eucalyptus species*
nimb: *Azadirachta indica*, neem
nirmali: *Strychnos potatorum*
palak: *Spinacia oleracea*
palas: *Butea monosperma*
pan: *Piper betle*, betel-leaf
pipal: *Ficus religiosa*
rai: *Brassica juncea*
ritha: *Sapindus trifoliatus*
rui: *Calotropis gigantea*
saag: *Tectona grandis*, teak
sabar: *Euphorbia neriifolia*
sagargota: *Caesalpinnia crista*
savaar: *Bombax ceiba*, silkcotton
sher: *Euphorbia tirucalli*
shevga: *Moringa oleifera*
shikakai: *Acacia sinuata*
shimti: *Lannea coromandelica*
shindi: *Phoenix sylvestris*, wild date palm
shisham: *Dalbergia latifolia*, rosewood
sitaphal: *Anonna squamosa*, custard apple
supari: *Areca catechu*
suran: *Amorphophallus campanulatus*
tag: *Crotalaria juncea*, sann hemp
temburni: *Diospyros melanoxylon*
terda: *Impatiens balsamina*, balsam
til: *Sesamum indicum*, sesame
tulsi: *Ocimum sanctum*
tur: *Cajanus cajan*, pigeon pea
udid: *Vigna mungo*
val: *Dolichos lablab*
vari: *Panicum milliaceum*, common millet
varuna: *Crataeva nurvula*
vekhand: *Acorus calamus*
vikhari takla: *Tephrosia purpurea*
wad: *Ficus bengalensis*, banyan
warras: *Heterophragma quadriloculare*

Index

acid rain, 77, 81, 133
Adivasi Mahamandal, 105
Adivasis, xii, xiii, 10, 27–60, 98, 104, 105, 117, 132, 135, 148, 149, 165, 176, 183, 184
adulsa, 119
afforestation, 10
Agni Purana, 4
agriculture, 11–13
ahimsa, 130, 143
AIDS, 158
alcohol, 30, 197
allergy, 157, 159, 167, 173
allopathic medicine, 46, 147, 154, 160, 170, 179, 180, 207, 199
allopathy, 153–65
Alpana Pharma, 157
ambadi, 52, 205
anabaena, 119
analgin, 155–6
animal husbandry, Warli, 43–5
Annusandan, viii
antibiotics, 157, 159
Aryabhata, 18
Asian Vegetable Research and Development Center (AVRDC), 110
aspirin, 157
astronomy, 17–18
avidya, 149
AVRDC, *see* Asian Vegetable Research and Development Center
ayurveda, 169, 176, 174, 179, 180
ayurvedic system, 19, 100, 172
azadirachtin, 125, 126, 127, 128
azolla, 119, 120

Badische Aniline Company, 94

bamboo, 55, 121, 139
Bata, 100–102
Baya devi, 177–8
"Bell and Lancaster" system, 14
bhagats, 153, 159, 176–80
Bhopal, 90, 95, 138
Binomial Theorem, 18
biogas, 52, 134, 197
biomass, 66
biotechnology, 95, 102, 157–8, 168
Bodhghat river project, 132
Brooke Bond India Ltd, 96, 137
Brown Plant Hopper (BPH), 115–6, 121
Brundtland Report, 62, 83, 78, 85, 87
bubonic plague, 154
Buddha, 216

cancer, 49, 94, 103, 148, 158, 160, 168, 173
cardiac surgery, 161
cattle, xii, 44, 57, 58, 105, 138, 150, 190–92
CFC's, 66, 80, 84, 140
Chandrakant, 43
charcoal, 20–21, 197, 205
Chernobyl, 134, 135
chlorine, 173
cocoa, 102
coffee plantations, 5
Colgate Palmolive (India) Ltd, 102
Conservator of Forests, 6, 7
constant capital, 70–71
cotton, 11, 12, 22–3, 109, 205
crop rotation, 11, 40, 118
crop sustainability, 112–23
crops, Warli, 41–2

Cyanamid India Ltd, 92

dams, 64, 132–3, 142, 165–6
dandhavan, 151
daru, 30
DDT, 66, 80, 93
Deccan, 194
Deep Ecologists, 217
deforestation, 8, 29, 51, 81, 132, 136, 137, 165, 166
dharma, 216
dhedhar, 119
dioxin, 140, 168
diptheria, 154
disease, 165–9
  tropical, 158
doctors, 158–60
drill plough, 11
drug manufacturers, 155–8
drugs, 155–8
  addictive, 156
  allopathic, 170
durva, 211

East India Company (EIC), 2–3, 89
East India Gum, 200
East Indian Railway, 3–4
education, 13–16, 135, 146, 149
Employment Guarantee Scheme, 58
energy, 64
erosion, 29, 40, 113, 188–9
ethnobotany, 151
existence values, 75–6
export policy, 142
exports, 140–42

farming systems, 109–29, 189
fertilizers, synthetic, xii, xiii, xiv, 79–80, 110, 111, 113, 118, 119, 138, 167
fertilizers, organic, 44, 125, 187, 150
fibres, 200
  rui, 205
  synthetic, 140
  use of by Warlis, 52–3
firewood, 51, 132, 165, 166
  tax, 9
food, additives, 167
  exports, 141
  ouput, 12
  processing, 110
foreign exchange, 141, 156
Forest Act, 7, 21
Forest Bill, 10
Forest Conservation Act, 10
Forest Department, 54
Forest Institute, 8
forest legislation 55
forest produce, 54–6, 132
forestry, 142, 182–92
  formal, 182–4
  natural, 184–92
  scheme, 58
forests, 64, 70, 74, 83, 169
  conservation, 4–13
  neem, 199
  pests, 185–6
fuel, 120, 197
  rui, 205

Gaia hypothesis, 81
Gandhi, Mahatma, xiv, 13, 130, 131, 142, 150, 174
Ganesha, 211
GATT, *see* General Agreement on Tariffs and Trade
Gauri, 212
General Agreement on Tariffs and Trade (GATT), 23
Generalized System of Preferences (GSP), 23
garwal, 175
ghanis, 113–4
global warming, 39, 61, 73, 75, 77, 80
grass, 104–6, 112
green consumerism, 81–2
green fodder, 11, 138
greenhouse effect, 80
green manure, 40, 118, 197, 202
GSP, *see* Generalized System of Preferences

Hanuman, 211
Harappans, 7, 11, 16, 22, 143
Hatch Bill, 156
hazardous wastes, 135
health, Warli, 53–4
heavy-metal pollution, 140
herbal medicines, 44–5, 53, 100, 147, 150, 174, 178

226 ASKING THE EARTH

HHL, *see* Hindustan Lever Ltd
high technology, 90
High Input Varieties (HIVs), 39–40, 116, 117, 118, 123
high-technology equipment, 160
"high-tech" therapy, 160–62
High Yield Varieties (HYVs), 13, 39, 115
Himalaya Drug Co., 157
Hindustan Cocoa Products, 102
Hindustan Lever Ltd (HHL), 44, 59, 96–9, 111, 142
Hindustan Petroleum, 93
HIVs, *see* High Input Varieties
Hoechst, 95–6, 100, 155–7
holistic development, 169–80
holistic system, 148
Hoffmann-LaRoche, 156
Home Charges, 4, 12
Howard, Albert, 11–12
housing, Warli, 50–51
hunting and gathering, Warli, 45–50
hydro power, 64
hyperconsumption, 75
HYVs, *see* High Yield Varieties

IMF, *see* International Monetary Fund
immune system, 170
incremental environmental benefits, 77
Indigo Commission, 94
industrial toxins, 167
industry, 24–5
information monoculture, 144
information overload, 144
innoculation, 19
insect parasites, 120
insecticides, 93, 116, 123, 173
insulin, 158
Integrated Pest Management (IPM), 128
intercropping, 128
intergenerational equity 63, 67–8, 71
intergenerational liabilities, 68–70
International Monetary Fund (IMF), 142
International Rice Research Institute (IRRI), 40, 115–6
intragenerational equity, 62–3, 67, 72
intragenerational liabilities, 72–3
IPM, *see* Integrated Pest Management

irrigation, xi, 11, 12, 133, 137, 142
Iron Pillar, 21
IRRI, *see* International Rice Research Institute

Jains, 132, 175, 216
jiva-ajiva, 216

kahandol, 132
Kalliyug, 17
karandi, 46, 183
karvi, 50
knowledge, 144–52
Khadi and Village Industries Commission (KVIC), 98
Konkan, 40, 103–4, 117, 128, 184, 192
korphad, 179
Krishna, 208, 210
kubabul, 191–2

land revenue, 12
Lever, 96–9
life expectancy, 168
lindane, 93
Lipton, 97, 137
literacy, 13, 15, 38, 146

Mahabharata, 214
Maharashtra, 91, 99, 137, 214
Maharashtra Forest Development Corporation, 182–3
Maharashtra Prabodhan Seva Mandal (MPSM), ix, xi, 148
mahua, 30, 46, 120
malaria, 154, 199
malathion, 93
malnutrition, 154, 166, 169, 170
mango, 150, 177, 141, 212
manure, green, 40, 118, 202
 neem, 197–8
 organic, xiv, 12, 128
medicine, 19–20
 allopathic, 46, 150, 154, 160, 170
 herbal, 53, 100, 150, 147, 174, 178
 neem, 198
 rui, 203–4
 tulsi, 207–8
meliantriol, 125
mercury, 140
milk, 93, 104, 105, 114, 137–8, 172

mosquitoes, 93, 120, 128, 173, 208
MPSM, *see* Maharashtra Prabodhan Seva Mandal
multinational malpractices, 102
mundi, 118

Napier, John, 18
Narmada Valley river project, 10, 132, 165–6, 175
National Wasteland Development Board, 10
Natural Products Research Centre, 100
neem, 54, 103, 117, 118, 122, 123–8, 132, 177, 178, 179, 180, 193–201, 211
  commercialization, 128
  cultivation, 194–5
  economic importance, 201
  for pest control, 124–6
  uses, 105–201
non-renewable resources, 71
Novalgin, 155–6
nuclear energy, 64, 73, 134
nutrition, 168, 169, 170–73
  unhealthy practices, 171–2

oilseeds, 54, 109, 113, 114, 189
Operation Flood (OF), 138
opium, 136
organ transplants, 161–2
organic manuring, 128
ozone-depleting chemicals, 78
ozone layer, 75, 80

Pancha Mahabhutas, 214
Pareto optimality, 68, 72
PCB, 66, 81
Pearce Report, viii, 62–88
PHC, *see* Public Health Care Centre
PMP, *see* Pollution-Multiplied Population
Pollution-Multiplied Population (PMP), 83–4
Proctor and Gamble, 99–100
Public Health Centre (PHC), 162–3
PVC, 140
paddy, 41–2, 117, 119, 120, 121, 150
  cultivation, 117–23
  threshing, 106–7
paracetemol, 156

pesticides, plant, xiv, 123–6, 150
pesticides, synthetic, xii, 13, 49, 91–4, 110, 111, 120, 123, 124, 128, 135, 167, 173
pests, 123–6
plant pesticides, 123–6, 150
plastics, 139, 140
pollution, 58, 61, 64, 66, 68–9, 75, 81–2, 91, 101, 113, 134, 135, 140, 141, 166, 168, 215
pollution, heavy metal, 140
  thermal, 133
  transborder, 78
pollution control, 90
population, 83–4
poverty, 169
preventative health care, 169
public health services, 162–3

Richardson Hindusthan, 100
Richaria, R.H., 117
Rita, 215
rab, 117
radioactive wastes, 64, 134
radioactivity, 134, 135
rainfall maintenance, 188
reclaiming salt lands, 202–3
recycling, 64, 66, 112, 139, 187
reforestation, 188
  programmes, 56
renewable energy, 134
rent-free land, 15
resource depletion, 69
rice, 3, 42, 118, 121, 141
  hybrid, 117
rui, 117, 121, 201–6, 211
  uses, 202–6
rui fibre, 205

sacred groves, 210
sal seeds, 98
salannian, 125
sanitation, 154, 168, 169, 173–4
Sardar Sarovar, 10
sarpagandha, 147
seeds, 56, 110, 118, 189, 190, 194, 196
Sesbania, 192
Shiva, 212
Shrinath Industrial Estate, 57–8

silk, 2, 22–3
smallpox, 19, 177
sodium nimbinate, 199
sodium niumbidinate, 199
solar energy, 64, 73, 132–4
state violence, 135–6
steel, 20–21
steroids, 157
subabul, 192
subsidies, 142
sugar, 137
sugar-cane, 137
superstition, 146, 147, 148
surgery, 20, 161, 179–80, 207
Sushruta, 147, 173
Sushruta Samhita, 147
sustainable civilization, 148
sustainable development, 61–3, 67, 69, 72, 75, 79
swadeshi, 174
synthetic fertilizers, xii, 13, 40, 66, 79–80, 111, 113, 118, 119, 138, 167
synthetic fibre, 140
synthetic pesticides, xii, 13, 13, 66, 91, 111, 120, 123, 124, 128, 135, 167, 173

tadi, 30
tag, 119, 122
Tata Tea, 137
taxation, 15
tea, 97, 136–7
tea trade, 136
teak, 6, 132, 183
technology, 16–24, 68, 147, 148, 150, 152, 161, 168
Tehri river project, 132
terda, 211
textiles, 22–24, 94
thalidomide, 155
timber, 5, 6, 72, 113, 187
 neem, 200
TNCs, *see* transnational corporations
tomatoes, 109–12
toxic chemicals, 61, 81, 215
toxic wastes, 135
transfer pricing, 82
transnational corporations (TNCs), 89–91, 95–103, 109, 111, 137, 138, 158, 166, 167, 175

tuberculosis, 154, 159, 162, 199
tulsi, 128, 173, 206–9, 210
 as medicine, 207–8
 uses, 207–9
Two-Thirds World, vii, viii, 68, 69, 76, 78, 83, 84, 85, 90, 100, 114, 155, 175
typhoid, 154

Unilever Group, 96–7, 98, 99
Union Carbide, 91, 93
University of Nuddea, 14–15
udid, 122
unnatural selection, 157

Valium, 156
vegetables, 112, 141, 173
 hybrid, 111
vidya, 149
vihkari takla, 118
Village Health Workers, 163, 176
violence, 130–43, 215
 permissible, 142
 sources of, 131–42
Vishnu, 208, 210

Warlis, xiii, 27–60, 151
 agricultural system, 39
 animal husbandry, 43–5
 crops, 41–2
 culture, 33–5
 forced labour, 32
 health, 53–4
 housing, 50–51
 hunting and gathering, 45–50
 knowledge, 36–9
 land ownership, 31
 taxation of, 30–32
wad, 211
wastes, radioactive, 64, 134
 toxic, 135
water, xii, 42–3, 138, 142, 166, 169, 173–4
WHO, *see* World Health Organization
wind energy, 64, 134
wootz, 20–21
World Bank, 142, 166, 183
World Health Organization (WHO), 91, 93, 156, 163, 169